iPhone X Manual for Beginners

iOS 11

The Perfect iPhone X Guide for Seniors, Beginners, & New iPhone X Users

Joe Malacina

TABLE OF CONTENTS

INTRODUCTION

Congratulations! So you have decided to take the first step, in fact the only step needed to learn how to use the iPhone X. Maybe you do not even have an iPhone X yet, and just want to see how it works before you decide whether to buy one. Either way, this book will teach you everything you need to know on using your iPhone. I wanted to take this time to tell you how this book is going to be the only tool you will ever need in order to learn your iPhone X. You see, this book was written for the perspective of a beginner. In other words, if you have never used a "smartphone" in your life, that will be no detriment when reading this book. That is the big difference between this book and other competitors available. Many authors fail to realize that even in 2018, many people are buying their first "smartphone," and need to be shown from the ground up the basics of using and navigating their device. With the iPhone X, it is especially important to understand the basics, as this particular phone works like no other iPhone. So that is what this book sets out to accomplish. I teach you not only how to do specific functions on your iPhone X, but I will teach you the building blocks of using any "smart" device. When you are finished reading this book, not only will you be a pro using your iPhone, you will be able to pick up any smartphone or tablet and have a general understanding of how it works and how to accomplish tasks.

This book is structured so that basic concepts I teach you in beginning chapters will be used in later chapters. Therefore, I highly recommend reading the first few chapters, instead of skipping ahead to exactly what you want to learn. You may miss out on essential tidbits of information that I will not cover in detail in later chapters.

Lastly, you may be wondering if this book is suitable for you. Will it answer all the questions you have? Will you understand the material? I can assure you that the information addressed in this book is derived directly from the input of several thousand iPhone users who have had the same questions as you. I have been teaching people how to use their iPhone for over 4 years, and have taught well over 50,000 people of all ages and backgrounds how to use their devices. I have run a blog and numerous websites where I have received over 10,000 emails with questions on how to do this and how to do that. So I can assure you, I know what the most common questions are, and where the most confusion lies. Take solace in the fact that this book will address your questions with a step-by-step approach, while building your technological intuition. When you are done, you will not even need to memorize the steps to perform a function, you will have the intuition and knowledge to figure it out quickly. That is the core of what this book will teach you.

On a final note, Apple users, particularly iPhone users, are notoriously loyal to their cell phones, and tend to keep using iPhones each coming year. After reading this book, you may start to see exactly why. The iPhone interface and navigation is seamless and soon you will find yourself liking the way it does things. Soon after that, you will notice that many other Apple devices work in very similar ways, and may find yourself wanting to explore other Apple devices such an iPad or Mac. If you ever find yourself wondering how these devices work, I urge you to check out www.infinityguides.com. There you can find guides on many Apple devices.

On that introduction, let us get started.

Chapter 1 – About the iPhone X

So you now have an iPhone X and you're ready to start using it. So what is a new iPhone X exactly? Your iPhone is classified as a smartphone, which means that it can do everything a cell phone can do, plus additional features. So with your iPhone X, you can make and receive calls, browse the internet, send text messages, check your email, and use apps. You can do all this with a phone that completely utilizes a touch screen. Before I show you how to do all these tasks, let's familiarize ourselves with some key terms that I will use often in this book. These terms are important for you to remember, as you will encounter them often.

Key Terms

Apple ID – Let's get the trickiest one out of the way early. By far, the most common question I receive is about the Apple ID. Let me define exactly what an Apple ID is. If you have ever used an Apple device before, such as a Mac computer or iPad, you have probably encountered this term before. An Apple ID is just an email address that is associated with you and your Apple devices. You need only one Apple ID, and you can use that same Apple ID on all of your Apple devices. I will go into creating an Apple ID in a chapter coming up, but for now just know that your Apple ID is your account that you will use on your iPhone. This account saves all of your important data such as contacts, photos, and your purchases.

Apps – Apps are programs on your iPhone that can do tasks. Your iPhone comes with many apps already installed, and you can download many more through the App Store. Nearly every aspect within the iPhone is part of an app, which you will see later on. Apps appear as square icons on your iPhone's home screen. (See Figure 1.1).

iTunes – iTunes is an app and a piece of software that is used to help you manage your iPhone's settings. There is an app on your iPhone called iTunes, where you can purchase and download music. More importantly, you can and should download iTunes on one of your computers. iTunes software on a computer backs up your iPhone's data every time you plug it in, and iTunes can be used to reset your iPhone in case there is a major problem with it. It is not necessary and you do not need to do this now, but I highly recommend that at some point you download and install iTunes on one of your computers. Instructions on how to download iTunes can be found on the website www.applevideoguides.com/itunes-install.html.

Home Screen – Throughout this book, you will see the term home screen used often. Your home screen is the main screen of your iPhone where all of your Apps are shown. (Figure 1.1)

Portrait & Landscape – You can view your iPhone in portrait or landscape mode. Portrait mode is the standard mode as shown in Figure 1.1. Landscape mode is when you turn your iPhone horizontal, and the contents of your iPhone also becomes horizontally oriented. Many apps allow you to use your iPhone in landscape mode and some require it. Just know that you can change the orientation of your iPhone, and many apps allow you to do so.

Different iPhone Models Explained

There are many different iPhones available in the marketplace, and it is important to understand the main differences between them. Generally, most iPhones work almost entirely in the same manner. For instance, performing most tasks and functions on an iPhone 5s can be done in the exact same manner on an iPhone 8. The reason this is important is so you understand that all iPhones operate nearly exactly the same, depending upon which software they are using. So if one friend has an iPhone 8, and your other friend has an iPhone 6, they will operate in almost the exact same way in every single aspect as long as both iPhones are using the same software version. The main differences between iPhones are where some buttons are located, such as the power button, the hardware underneath the iPhone, and with the iPhone X, the lack of a *home button*. Some older iPhones will have different charging ports as well. All in all, remember this: No matter which iPhone you have, it will work nearly the same as every other iPhone that is running the same software.

Let's now explain the software running on iPhones. The software your iPhone is running is called its iOS. iOS stands for I Operating System, and it governs how everything in your iPhone works. iOS is labeled by a number, and the higher the number, the newer the software. For instance, the iPhone X comes with iOS 11. The iPhone 6S originally came with iOS 9. At all times you should be using the newest version of iOS available for your iPhone. I will show you how to update the iOS software in Chapter 3.

All of these square icons are apps

Figure 1.1 – The Home Screen

Why the iPhone X is Different

The iPhone X, pronounced "iPhone ten", is the first iPhone of its kind to do away with the signature home button. On previous iPhones, the home button was the circular button at the bottom on the front of the device. This home button was essential to using the iPhone as it would always bring you back to your home screen when pressed, and also performed some other useful functions. Now with the iPhone X the home button is no more, and has instead been replaced with the **home swipe**, which is a gesture that essentially performs what the home button still does on other iPhones. If you have used iPhones in the past, adjusting to this *home swipe* is going to take some time. This book will get you accustomed to using the *home swipe* and we will explain everything along the way; and we start now in Chapter 2.

Chapter 2 – iPhone X Layout

Now we enter the "instruction manual" portion of this book, and we start with the iPhone layout. Shown in <u>Figure 2.1</u> is the layout of an iPhone X.

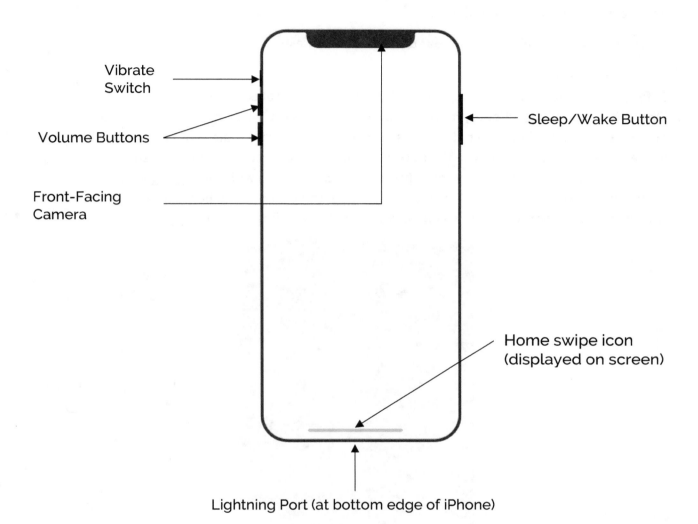

Vibrate Switch

Volume Buttons

Front-Facing Camera

Sleep/Wake Button

Home swipe icon (displayed on screen)

Lightning Port (at bottom edge of iPhone)

<u>Figure 2.1</u> – Layout of iPhone X

As you can see, the sleep/wake button (also known as the power button) is on the right side of the phone, while the volume buttons are on the left. Pressing the <u>sleep/wake button</u> turns the iPhone's screen on or off. In other words, pressing this button either puts your iPhone to sleep or wakes it up; it does not completely turn the iPhone off. The volume buttons are used to control the volume of the iPhone. Pressing the <u>top volume button</u> increases the iPhone volume, while pressing the <u>bottom volume button</u> decreases volume. Above the volume buttons is the <u>vibrate switch</u>. When the switch is flipped up, sound mode is on for the iPhone, which means your iPhone will ring when called and play sounds. When you flip the vibrate switch down, this turns on vibrate mode. In this mode, the iPhone will not ring, and will not make sounds.

At the bottom of your iPhone is the charging port. This port is called the lightning connector port, and allows you to charge your iPhone, connect accessories such as headphones, and connect your iPhone to other devices.

On the back of your iPhone is the rear camera and the flash.

The Home Swipe

Before getting into anything else, we need to cover what is called the *home swipe* on the iPhone X. On previous iPhones, there was a circular button at the bottom of the front of the device. Pressing this button would always bring you back to your main screen, known as the home screen. You could also use the button to turn on your iPhone's screen when it was asleep. With the iPhone X, the home button has been replaced with the home swipe, and it is pretty simple to perform. To perform a home swipe: **tap down at the bottom of your screen with your thumb or finger, and swipe up and release your finger** (Figure 2.2).

To perform a home swipe, tap down at the bottom of your screen with a finger or thumb, and swipe it up the screen releasing your finger off the screen at the end. Do this quickly in one fluid motion.

NOTE: You do not necessarily need to swipe your finger/thumb far upwards to perform a home swipe, although you can if you want to. A ½ inch or 1 centimeter swipe from the bottom will do the trick.

TIP: When holding your iPhone in your hand, it is easiest to use your thumb to perform a home swipe. When your iPhone is placed on a surface, it is usually easiest to use a finger for a home swipe.

Figure 2.2 – How to Perform a Home Swipe

You can do the home swipe at any time, including when your iPhone is in horizontal orientation. Practice this home swipe now as you will be performing it often. The best way to practice it is to first open any app by tapping on a square icon on your home screen (Figure 2.3.1); now perform a home swipe (Figure 2.3.2) and you should be brought back to your home screen. Try it out a couple of times to see how it works. This book will be instructing you to perform home swipes very often and will easily be the gesture you use the most on your iPhone X, so it is prudent to become familiar with it now. Each time we instruct you to perform a home swipe, it will be underlined in this text.

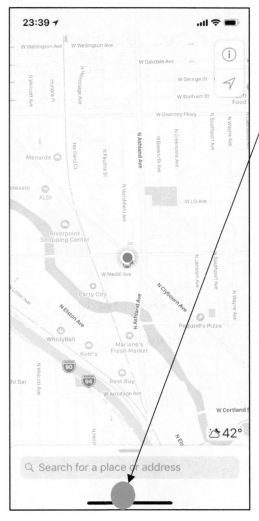

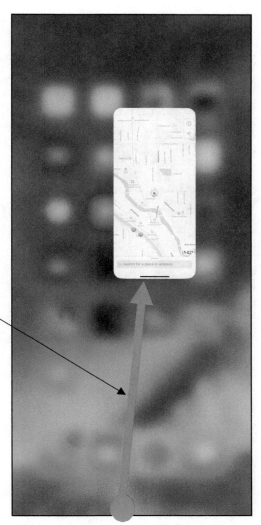

First, tap down at the very bottom of the screen (a line may appear for reference)

Then swipe up and release. Perform the home swipe quickly and in one fluid motion.

Figure 2.3.1 – Practice the Home Swipe inside an App (Maps app)

Figure 2.3.2 – The Home Swipe in Slow Motion

You can perform the home swipe at ANY TIME, and it will always bring you back to your home screen. You may need to perform the home swipe twice in order leave certain apps or when you are using your iPhone in landscape orientation.

Turning your iPhone X on and off

When your iPhone is off, you can turn it on by pressing and holding the <u>sleep/wake button</u> until your screen lights up. This will turn your iPhone on and it will start booting up. After several seconds it will boot up completely.

At any time you can turn your iPhone X off by pressing and ***holding*** the <u>sleep/wake button</u> and any <u>volume button</u> until <u>Figure 2.4</u> appears. To turn your iPhone off from this screen, place your finger down on the <u>red power symbol</u>, and slide it across to the right, releasing at the end. This turns your iPhone off.

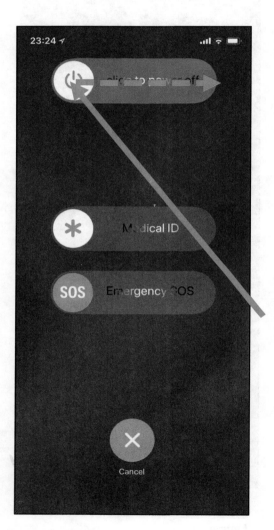

To turn iPhone off from this screen: Tap on the power icon with your finger, hold down, then swipe your finger to the right, releasing at the end.

<u>Figure 2.4</u> – **Power the iPhone X Completely Off**

When your iPhone X is on and the screen is asleep, you can wake the screen by either pressing the <u>sleep/wake button</u> or by tapping on your screen. The screen will also awaken when you lift your iPhone X to your face.

Charging your iPhone X

To charge your iPhone, plug the included USB cable into the bottom of your iPhone, and connect the opposite end (USB) into a USB port or charging dock. Any powered USB port will charge your iPhone including USB ports on your computer or vehicle. The charging dock can be plugged into an electrical outlet. Once your iPhone is fully charged, you should disconnect the charging cable. For optimal battery life, the best practice you can follow is letting your iPhone's battery drain to yellow or red levels, then charge it until it is full, followed by unplugging the charging cable.

You can also charge your iPhone X wirelessly by using a wireless charging pad. To do this, simply place the back of your iPhone X onto the charging pad, and it will begin to charge without you having to plug anything into the iPhone. Wireless charging pads are a separate accessory that are not currently included with the iPhone X and can be purchased separately.

Chapter 3 – Getting Started

This chapter covers getting started with your iPhone X, and goes over the first-time setup procedure. If you have already completed the first-time setup where you chose your language and created your Apple ID you can skip the first part of this chapter.

First-time Setup

When you power on your iPhone for the very first time (press and hold the <u>power button</u> when the iPhone is off), you will be brought to the first-time setup screen shown in <u>Figure 3.1</u>. This first screen will say Hello in multiple languages. Here are the steps required to move past the initial setup successfully.

1. On the Hello screen (<u>Figure 3.1</u>), tap down at the bottom of your screen and swipe up (<u>home swipe</u>) to move to the next screen.

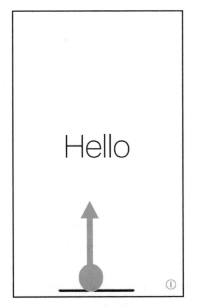

To move to the next screen, tap down on the line at the bottom of your screen with your finger, hold it there, and slide your finger upwards, and release your finger at the end. This is called a home swipe.

<u>Figure 3.1</u> – Hello Screen

2. You are now asked to select your language (<u>Figure 3.2</u>). Tap with your finger one time on the language you wish to use for your iPhone. If your language does not appear, tap down on the screen and move your finger up or down to scroll through the list of languages. Release your finger to complete the scroll. (Swiping your finger up and down to move through a screen is called **scrolling**, whereas tapping down on your screen with a finger and moving it in any direction and releasing is called a **swipe**.)

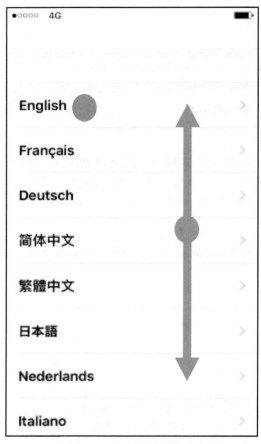

Figure 3.2 – Language Screen

To choose your language, you can tap down on the screen with your finger, and move your finger up and down the screen to "scroll" through different choices. When "scrolling", remove your finger at the end of each movement. Think of it as quickly flicking your finger up or down the screen. To choose your language, tap on the name of the language with your finger and release.

3. You will now be asked to select which country or region you live in. Select your country or region by tapping down on the correct one using your finger and release. Again, you can scroll up and down by tapping down and moving your finger up or down and releasing to move throughout the list.

4. The next screen is called the Quick Start screen, and allows you to quickly setup your iPhone X if you have any other iPhone or iPad around that is running iOS 11 or newer (software version). If you do have one and want to set up your iPhone this way, just turn on the other device and bring it next to your iPhone X. Doing so will set up your iPhone X with the same settings and Apple ID as your other device. If not, or if you prefer to set up your iPhone X manually, tap on the text at the bottom that says Set Up Manually. (In this book, we will be setting up the iPhone X manually.)

5. The next screen will ask you to choose your Wi-Fi network. It is best to setup your iPhone at home so you can connect to your home wireless network. After a few moments, a list of available networks will appear on your screen. Find your home network and tap on it with your finger. A new screen will appear asking for your wireless network password. Use the keyboard that appears at the bottom of your screen to tap in the password. Simply tap each letter or number one at a time to enter

it in. You can use the "<u>back arrow</u>" key on the keyboard to backspace, and you can use the "<u>123</u>" key to see numbers and symbols. Furthermore, to capitalize letters, you can use the <u>shift key</u>, which is the key with the upward arrow. Tapping on it will make the letters capital for ONE entry. Once you have entered your Wi-Fi password tap <u>Join</u> on your screen. (See <u>Figure 3.3</u>)

    a. If you do not have Wi-Fi available near you, or are having trouble connecting to your Wi-Fi network, you can tap on <u>Use Mobile Connection</u> at the bottom of this screen.

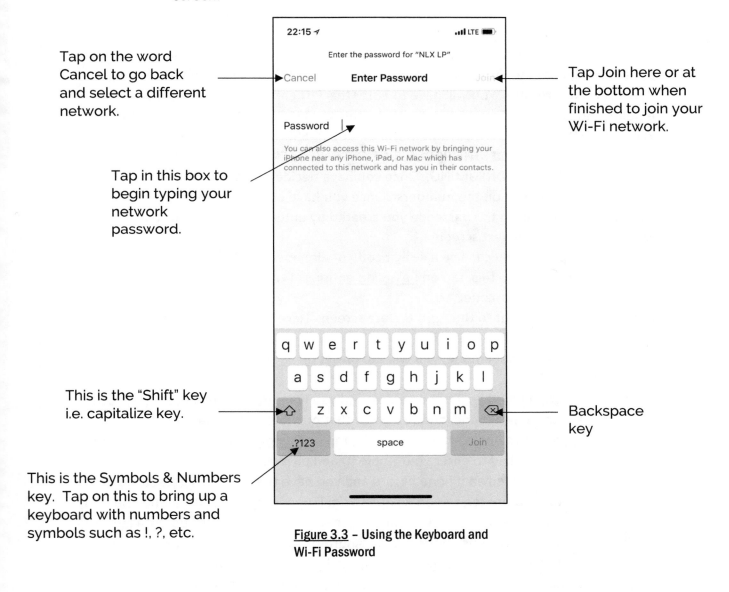

Tap on the word Cancel to go back and select a different network.

Tap Join here or at the bottom when finished to join your Wi-Fi network.

Tap in this box to begin typing your network password.

This is the "Shift" key i.e. capitalize key.

Backspace key

This is the Symbols & Numbers key. Tap on this to bring up a keyboard with numbers and symbols such as !, ?, etc.

<u>Figure 3.3</u> – Using the Keyboard and Wi-Fi Password

6. Now your iPhone X will attempt to activate itself. This can take a few minutes. Once this is completed you will be brought to a new screen to set up Face ID.

7. Now you will be prompted to set up Face ID. Face ID allows you to unlock your iPhone just by looking at it. I recommend setting this up now. Tap on <u>Continue</u>.
   a. Tap <u>Get Started</u> and follow the instructions on your screen to setup Face ID. You will be asked to hold your iPhone X up and look directly at the screen, bringing your face into position inside the circle. Once your face is positioned, hold your iPhone steady and move your head around in a complete circle.
   b. Once the first scan is complete, tap <u>Continue</u>.
   c. Complete the circular motion again.
   d. Once the second scan is complete, your iPhone X will show "Face ID is now set up." Tap <u>Continue</u>.
8. Next, you will be asked to setup a passcode. A passcode is a 6-digit PIN that allows you to unlock your iPhone. You will need to remember this PIN, so I recommend writing it down somewhere safe. You will have to enter this PIN each time you restart your iPhone, and each time Face ID fails to work properly. I CANNOT STRESS ENOUGH THAT YOU SHOULD WRITE DOWN AND REMEMBER THE PASSCODE YOU CREATE. IF YOU FORGET YOUR PASSCODE, THE PROCESS OF GETTING BACK INTO YOUR IPHONE IS DIFFICULT AND TIME CONSUMING. Once you have decided on a passcode, enter it in your iPhone by tapping on the numbers. Once you have entered it the first time, you will be asked to confirm the passcode you created by entering it again. Do so and you will be brought to the next screen.
   a. If you prefer, you can use a 4-digit code or alphabetic code instead of a 6-digit passcode. To do this, tap on <u>Passcode options</u>. I will also show you later how to change your passcode.
9. Next you will be brought to the Apps & Data screen. Here you have to choose how to setup your iPhone.
   a. If this is your first iPhone, you will tap on <u>set up as new iPhone</u>. **Most Likely Option.
   b. If you have owned an iPhone before, and you want to migrate all your previous data to this new iPhone, and you know that previous data is backed up on iCloud, which is Apple's backup service, tap on <u>Restore from iCloud Backup</u>. You will then be asked to enter your Apple ID and password.
   c. If you have owned an iPhone before and you have that iPhone's data saved on iTunes on a computer, tap on <u>Restore from iTunes Backup</u>. You will then be told to plug your iPhone into your computer that has the iTunes backup installed.
   d. Lastly, if your previous phone was an Android phone and you want to migrate that data over, tap on <u>Move Data from Android</u> and follow the instructions on the screen. This option is a little more complicated and not recommended for new users.
10. Next, you will be brought to the Apple ID screen. THIS SCREEN AND THIS STEP IS VERY IMPORTANT. PLEASE FOLLOW THIS STEP VERY CAREFULLY. On this screen you are asked to login with your Apple ID. If you have ever owned an iPhone or iPad, chances

are you have an Apple ID. Furthermore, if you currently use a Mac computer, you may already have an Apple ID as well. This next sentence is very important. If you have other Apple devices and have an Apple ID, it is very important that you use the same Apple ID for all of your Apple devices. This way, all of your purchases on one Apple ID will be available on all of your devices. Just to reiterate from earlier, an Apple ID is an email address you have. If you know you already have an Apple ID, go to step 10A. If you do not have an Apple ID or are unsure, go to step 10B.

a. If you already have an Apple ID from previous Apple devices, tap into the box to the right of the text "Apple ID" and enter your Apple ID email address. Remember, this is your email address that you have registered as an Apple ID with Apple. After you enter your email address, tap into the box next to password and type in your password using the keyboard that appears. When you are done, tap <u>Next</u> at the upper right.

b. If you do not have an Apple ID, or this is your first Apple device, or you are not sure if you have an Apple ID, tap on the text at the bottom that says "<u>Don't have an Apple ID or forgot it?</u>"

    i. On the next screen you can tap on "<u>Forgot Apple ID or Password</u>" if you think you have an Apple ID from a previous Apple device. If you do not have an Apple ID, tap on <u>Create a Free Apple ID</u>. **PLEASE NOTE, SETTING UP AN APPLE ID IS HIGHLY RECOMMENDED, AS YOU NEED AN APPLE ID TO DOWNLOAD APPS, BACKUP YOUR DEVICE, AND USE OTHER CRITICAL FEATURES.

        1. First, you will be prompted to enter your birthday. Do so by tapping in each area at the bottom of your screen and scrolling up or down to select your month, day, and year. Tap <u>Next</u> at the upper right when done.

        2. Now you will be asked to enter your first and last name. Tap into each corresponding box and enter your name. Tap <u>Next</u> at the upper right when done.

        3. Now you will have to enter in some information to create your Apple ID.

            a. Your iPhone will now ask which email address you want to use as your Apple ID. You can either use an email address you already have or create a new iCloud email address. I suggest using an email address you already have. Tap <u>Use your current email address</u>.

            b. In the email box, tap into the box next to email and type in the email address you want to use as your Apple ID. I suggest using your personal email address that you use fairly often. Then tap <u>Next</u> at the upper right.

    c.   Next you must create a password for your Apple ID. This box is NOT asking for your email password, it is simply asking you to create a password for your Apple ID. Tap into the password box and type in a password for your Apple ID. The password must contain at least 8 characters, and it must contain at least 1 capital letter, 1 lower case letter, and at least 1 number.

    d.   Now tap into the verify box and type in your new password again to verify it is the same. REMEMBER THIS PASSWORD. WRITE IT DOWN SOMEWHERE SAFE IF YOU MUST, BUT YOU WILL NEED THIS PASSWORD LATER. When done, tap <u>Next</u> at the upper right.

    e.   Next you will need to verify your identity using a phone number. The phone number of your iPhone X should automatically fill this box, so just tap on <u>Next</u> at the upper right here. If your phone number does not fill in automatically, enter a phone number you can be contacted at.

    f.   Your iPhone X should automatically verify your identity at this point. If you had to enter a phone number that is not the number of your iPhone X, that phone will receive a text message or a phone call with a verification code. Get the code and enter it in on the next screen if applicable.

    g.   Next will be the Terms and Conditions screen. Read these and tap on <u>Agree</u>, if of course you do agree to these terms and conditions, at the bottom right to proceed.

    h.   Your Apple ID is now being created. This can take a few minutes and once it is done you will be brought to a screen that will prompt you to setup your email address that you used for your Apple ID on your iPhone. This step is completely optional. Tap <u>Next</u> at the upper right.

11. The creation of your Apple ID is over. Now you will be brought to an Express Settings page. Tap on <u>Continue</u> at the bottom to proceed.

12. The next screen will be the Siri screen. On this screen you will be asked whether you want to setup Siri. I recommend you set it up now, so tap on <u>Continue</u>. You will be asked to speak a few phrases at your iPhone. Follow the directions that appear on your screen. You will have to speak about 5 phrases. All this does is train Siri to recognize your voice. Siri is simply Apple's artificial helper, which listens to you and follows your command. Once this step is complete, your iPhone will read "Hey Siri is Ready," and you can tap on <u>Continue</u>. If you are having trouble setting up Siri, just tap on <u>Set Up Siri Later</u> at the bottom of your screen.

13. The next screen will be the iCloud Analytics screen, which is basically asking you whether you want your iPhone to periodically send data to Apple to help Apple fix bugs and glitches. This is completely up to you if you want to enable this or not. Tap on <u>Share with Apple</u> or <u>Don't Share</u> to continue.

14. Next a screen with "App Analytics" may appear. This is similar to the previous step. Tap on <u>Share</u> or <u>Don't Share</u>.

15. The next screen will tell you about True Tone Display. Read the information then tap <u>Continue</u>.

16. The next screen will teach you about the home swipe (covered in Chapter 2). Tap <u>Continue</u>.

17. The next screen will teach you about the <u>App Switcher</u> (covered in Chapter 21). Tap <u>Continue</u>.

18. The next screen will teach you about accessing the Control Center (covered in Chapter 18). Tap <u>Continue</u>.

19. You are all done! You should see the "Welcome to iPhone" screen. Perform a <u>home swipe</u> to proceed and to access your home screen. (Remember, a home swipe is tapping down at the bottom of your screen and swiping up and releasing your thumb or finger.)

After you have completed the initial setup you will be brought to your home screen. Remember, your home screen is the screen that shows all your apps.

Chapter 3.5 – Update your iOS Now

Before we begin Chapter 4, it is important that you check to see if an update is available to your iPhone's software, otherwise known as iOS. In order to do this, follow these steps:

1. On your home screen, which shows all of your apps, there should be an app called Settings. Tap on the app with your finger to open it. (Figure 3.5.1)

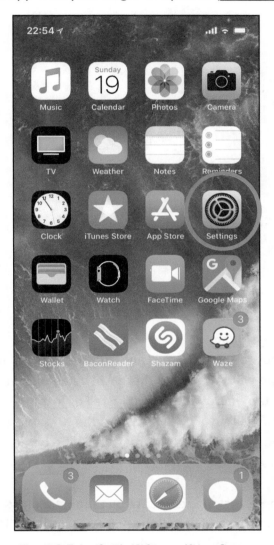

Figure 3.5.1 – Settings App on Home Screen

2. Next, tap on the word General, which you should see somewhere on your screen. You can scroll up and down if you cannot find it.

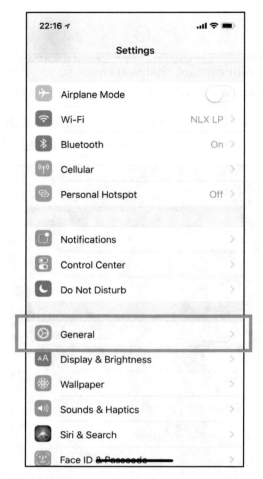

Figure 3.5.2 – Settings -> General

3. Next, tap on <u>Software Update</u>.
4. Your iPhone will check to see if there is an update available. If there is an update available, your iPhone will say so, and there will be an option to <u>Download and Install</u>. If this option appears, tap on <u>Download and Install</u> or <u>Install</u>. If there is not an update available, your iPhone will say iOS X.y.z, where the X, y, and z will be numbers. For the sake of this book, the only number that matters is the first number. So for this book, you will want that first number to be 11 or greater. It will also say your Software is Up to Date.
5. If an update is available, install the update by tapping on <u>Install</u> or <u>Download and Install</u>. Your iPhone will download and install the software update. This could take some time, and you will need to be connected to Wi-Fi to download the update. (See Chapter 4 to learn how to connect to Wi-Fi if you did not do so in the initial setup). Let the update install, and your iPhone will restart when it is finished, and then you are ready to continue. If your iOS software is already up to date, then you are ready to continue.
6. You can perform a <u>home swipe</u> on your iPhone to return to the home screen.

If a software update is available for your iPhone X, it is HIGHLY recommended that you update it.

Chapter 4 – Navigating your iPhone X

All you need to use your iPhone are your fingers. Everything is based on the touch screen and the gestures you can perform on the touch screen such as the home swipe. Here at the home screen (<u>Figure 4.1</u>), which is just the screen that shows you all of your apps, you can open an app, which is a program on your phone that can do tasks and enable features, by lightly touching down on the square with your finger and releasing quickly. To then leave the app and return to the home screen, perform a <u>home swipe</u>.

Your home screen is actually made up of multiple screens to accommodate however many apps you have on your iPhone. To move through different pages of apps on your iPhone's home screen, touch down on the screen with one of your fingers, and then drag your finger to the left or the right, like you would be dragging a page, then release. If at any time you need to return back to the main home screen, just perform a <u>home swipe</u>.

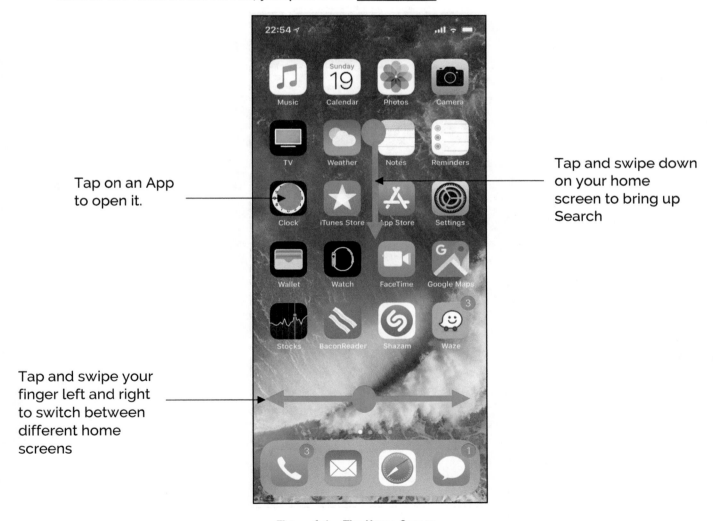

Tap on an App to open it.

Tap and swipe down on your home screen to bring up Search

Tap and swipe your finger left and right to switch between different home screens

<u>Figure 4.1</u> – The Home Screen

Search your iPhone

You can search your iPhone for anything using the Search function. To use this function, touch down somewhere on your home screen, and drag your finger down then release (Figure 4.1). This is a great way to search for apps on your iPhone when you have too many to remember them all. To get back to the home screen simply perform a home swipe, or swipe up with your finger.

Today Screen

You may have noticed, that when you are on your main home screen and you swipe your finger to the right that you are brought to what is called the "Today" screen (Figure 4.2). This screen shows you if you have anything coming up on your calendar, along with some suggestions for apps that you use. There may also be more information on this screen such as your local weather and news. You can tap on anything here to explore it more. To get back to the home screen perform a home swipe or swipe to the left with your finger.

Figure 4.2 – The Today Screen

Connecting to Wi-Fi

Connecting to Wi-Fi is one of the most important things you should do with your iPhone. When you are connected to Wi-Fi, your internet speeds are generally faster, and you are not using valuable cellular data. You may have connected to your home wireless network when you first set up your iPhone, in which case you are already good to go. You can tell if you are connected to Wi-Fi by looking for the Wi-Fi symbol at the top right of your screen. If the Wi-Fi symbol appears, it means you are connected to Wi-Fi (Figure 4.3). Once you have connected to a Wi-Fi network, your iPhone will remember that network, and you will not need to ever manually connect to it again. Each time your iPhone comes in range of that network it will connect automatically.

To connect to a Wi-Fi network, follow these steps:

1. Open the Settings app on your home screen by tapping down on it and quickly releasing.
2. Tap Wi-Fi
3. Your iPhone will search for networks. If you are already connected to a network, it will appear directly under the words Wi-Fi and a checkmark will be next to its name. Look at the network names under where it says "CHOOSE A NETWORK..." When you see the network you want to connect to, tap on it with your finger.
4. You will then have to enter your Wi-Fi password. Enter this password using the keyboard, and then tap on Join at the upper right. (For more on using the iPhone keyboard, see Figure 3.3 in Chapter 3).
5. Your iPhone will join the Wi-Fi network as long as the password was entered correctly. Remember, Wi-Fi passwords are case sensitive. Once you have joined a Wi-Fi network, you will never have to manually join that network again, as your iPhone will connect to it automatically from now on.
6. Perform a home swipe to return to the home screen.

TIP: If you do not know your home wireless network name and password, it may be located on the back or bottom of your wireless router. Check the back of your router to see if that information is there.

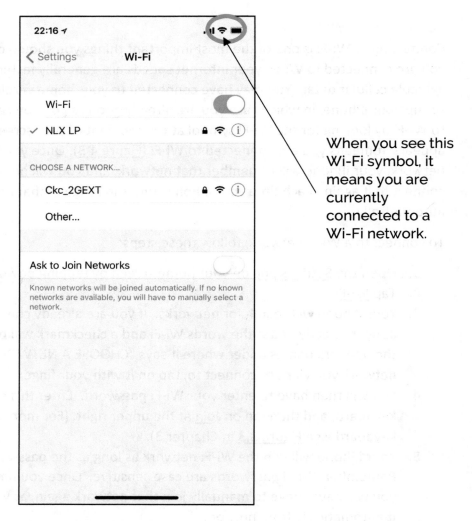

When you see this Wi-Fi symbol, it means you are currently connected to a Wi-Fi network.

Figure 4.3 – Wi-Fi

Chapter 5 – Apple ID

I have already discussed the concept of your Apple ID throughout this text, but it is worth emphasizing some more. The most common questions I receive about using iPhones are related to Apple IDs. So in this chapter, I will cover everything you need to know about your Apple ID.

How to Check Your Apple ID

If you created an Apple ID during the initial setup of your iPhone, great! If you did, please write down your Apple ID email and password somewhere safe. You will need that information from time to time. If you have not created an Apple ID or are not sure, there is a way to check. Follow these steps:

1. Open the <u>Settings app</u> on your iPhone by tapping on <u>Settings</u> on your home screen.
2. Scroll down through Settings until you find <u>iTunes & App Store</u> and then tap on it.
3. On this screen, at the top, will be a box labeled Apple ID: and inside this box will be your Apple ID if you have created one and are signed in. There should be an email address in there. If there is not an email address in there, it will read "Sign In" instead. (See <u>Figure 5.1</u>)

Now you know whether you are signed in to your iPhone with an Apple ID, and what that Apple ID is. If you are signed in, write down and remember your Apple ID. If you are not signed in, and do not have an Apple ID, you need to create one now.

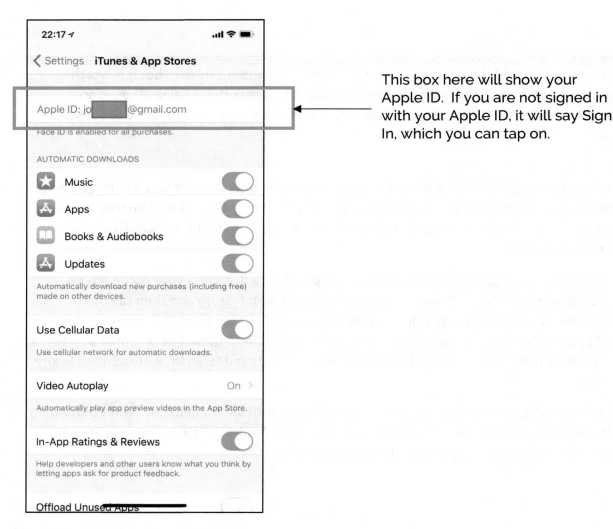

This box here will show your Apple ID. If you are not signed in with your Apple ID, it will say Sign In, which you can tap on.

<u>Figure 5.1</u> – Settings -> iTunes & App Store

How to Create an Apple ID

If you do not have an Apple ID, you need to create one. Simply put, you need an Apple ID to use many of the functions available on your iPhone. To create an Apple ID, follow these steps:

1. Go to your home screen. (Remember, you can get to your home screen at any time by performing a <u>home swipe</u>. If you forgot how to perform a home swipe, see Chapter 2.)

2. Find the app called <u>iTunes Store</u> and tap on it to open it.

3. When the app loads completely, scroll down to the very bottom and tap <u>Sign In</u>.

4. If you have an Apple ID, sign in with your Apple ID login credentials by tapping on <u>Use Existing Apple ID</u>. If you do not have one, tap on <u>Create New Apple ID</u>.

5. On the next screen, enter an existing email address that you use regularly. This email address will become your Apple ID.

6. Next you will have to create a password for your Apple ID. This is not the same as your

email's password. It is a separate password for your Apple ID. Type in a password and type it in again in the Verify box. Passwords must contain at least 8 characters, and must contain at least one number, and one capitalized letter and one lowercase letter.

7. Choose your region if it is not automatically filled in for you and agree to the terms and conditions (if applicable) by reading them and tapping on the tab next to them.
8. When done, tap <u>Next</u> at the upper right.
9. At the next screen you will have to fill in your personal information, and select security questions for your account. The security questions are for in case you forget your password. Fill everything in, along with the security questions and answers, and tap <u>Next</u> when done.
10. Next you will have to enter billing information. You will not be charged anything, but this billing information is required in case you decide to purchase anything in the future and to verify your identity. Fill out the form and then tap <u>Next</u> at the upper right.
11. There are a few more steps left that are pretty self-explanatory. Follow the instructions on your screen to complete the creation of your Apple ID. You may be asked to verify your phone number and/or your email address.
12. When all the steps have completed, your Apple ID will have been successfully created. Occasionally, you may be asked for your Apple ID password on your iPhone X. When you are asked, enter it in on your keyboard. Remember to use correct capitalization.

*Alternatively, you can create an Apple ID from a computer by visiting <u>https://appleid.apple.com/account</u> (Recommended if creating an Apple ID on your iPhone becomes too tedious or confusing).

Signing in with Your Apple ID

If you have created your Apple ID on your iPhone, you will most likely be signed in automatically on your iPhone. Your Apple ID is used for many apps, and it may be necessary to sign in to all of these apps with your Apple ID. Follow these steps to sign in to your iPhone with your Apple ID:

1. On your home screen open the <u>Settings app</u>.
2. Scroll down to <u>iTunes and App Store</u> and tap it.
3. Tap the Apple ID box at the top and tap <u>Sign In</u>. If your Apple ID already appears, you are already signed in.
4. Enter in your Apple ID email address and password and sign in.
5. Once signed in, tap the <u>back arrow</u> at the upper left where it says Settings.
6. Now tap <u>on the big box at the very top where it says Apple ID or your name</u> (<u>Figure 5.2</u>).
7. At the top will be your name and email address if you are already signed in to your Apple ID for iCloud. If your name does not appear, tap on <u>Sign In</u>.
8. Enter your Apple ID email address and password and tap <u>Sign In</u>.
9. Use the <u>back arrows</u> again at the upper left to return to the main Settings page.
10. Now we want to check and make sure your Apple ID is signed in for all Apple services.

11. Find in Settings the <u>Messages</u> box and tap it.
12. Now tap where it says <u>Send & Receive</u>. (<u>Figure 5.3</u>)
13. At the top where it says Apple ID, make sure your Apple ID appears there. If it does not, tap on it and sign in. You may also have to tap on <u>Use Apple ID for Messages</u>, it is appears.
14. Use the <u>back arrows</u> at the upper left to return to the main Settings page, and tap on <u>FaceTime</u>.
15. In the Apple ID box, make sure your Apple ID appears. If it does not appear, tap on the box and sign in with your Apple ID.
16. Perform the <u>home swipe</u> to return the home screen.

You should now be signed in completely with your Apple ID.

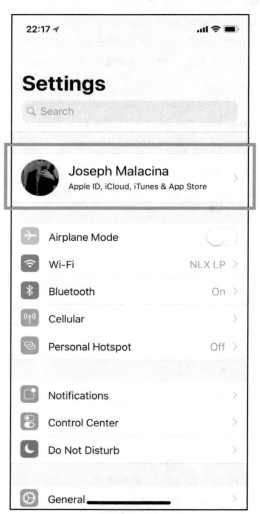

Figure 5.2 – Settings -> Big Box at Top (Apple ID)

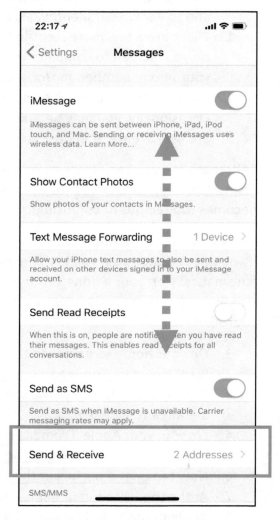

Figure 5.3 – Settings -> Messages -> Send & Receive

How to Use Your Apple ID

Now that you are all signed in with your Apple ID, how can you use it? You do not really need to worry about that actually. You will be using your Apple ID automatically. Every time you download an app you will be using your Apple ID. Occasionally, your iPhone may show a popup box asking you for your Apple ID password. When this appears, just enter your password and tap OK/Sign In. Your iPhone will ask for your Apple ID password periodically for security purposes. Make sure you remember this password!

All the confusing stuff is now out of the way. Once you are signed in with your Apple ID, you do not need to worry about it anymore as long as you remember your Apple ID password. It is finally time to get into actually using the iPhone.

Chapter 6 – Your Contact List

Your contact list is an essential tool on your iPhone X. With it, you can store your contacts' phone numbers, email addresses, and other vital data.

Importing Contacts

When you first get your new iPhone, you may want to transfer the contacts from your previous cell phone to your iPhone. The best way to do this is to have your wireless provider do it for you. Most wireless provider stores can transfer and import your contacts using their own equipment. Unfortunately, for some cell phones you will not be able to do this, in which case you will have to create your contacts manually.

If you have had an iPhone in the past, your contacts will transfer automatically when you sign in with the same Apple ID.   The same goes for Macs. If you created a contact list on a Mac computer, when you sign in with your Apple ID on your iPhone those same contacts will be automatically imported. You can also import your Outlook contacts to your iPhone using iTunes software. This is much more complicated, and can be learned by visiting www.applevideoguides.com and clicking on iPod & iTunes Guide at the top.

Creating Contacts

Let us see how we can create a contact. Here are the steps: (See Figures 6.1 & 6.2)

1. Open the Phone app from your home screen.
2. At the bottom, tap on the tab that says Contacts. (Figure 6.1)

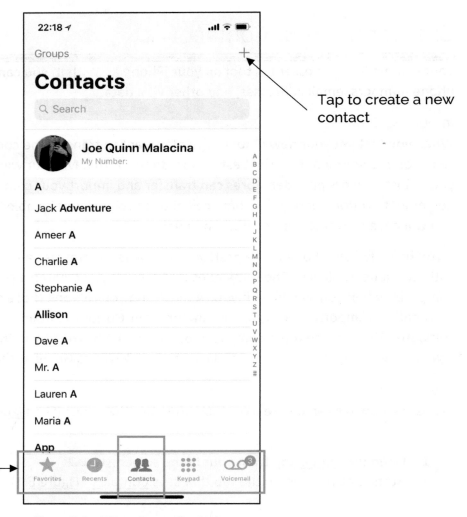

Tap to create a new contact

These icons at the bottom are called *tabs*. Tap on the <u>Contacts</u> tab to open the Contacts screen.

Figure 6.1 – Phone App -> Contacts

3. At the upper right of your screen, there will be a + sign. Tap on the <u>+</u> sign. (<u>Figure 6.1</u>)
4. You will now be brought to a screen where you can create a contact.
5. Tap into each corresponding box to type in an entry for the new contact. You can enter in their <u>First name</u>, <u>Last name</u>, and <u>Company</u>.
6. Tap into the <u>add phone</u> box to add their phone number.
7. Tap into the <u>add email</u> box to add their email, if applicable.
8. By scrolling down further, you can choose to enter in additional information including their <u>home address</u>, <u>birthday</u>, and other information.
9. You can also choose a specific ringtone for a contact.
10. Once you have added all the information you want for your new contact, tap <u>Done</u> at the upper right.
11. Your new contact has just been created.

Tapping on <u>add photo</u> will allow you to set a picture for your contact that you have saved in your Photos app.

When done tap here.

You can tap into each box to type in contact information.

Tap the green + icon to add the corresponding information such as a phone number or email address.

You can scroll (tap down on screen and swipe finger up/down) to enter more information for your contact.

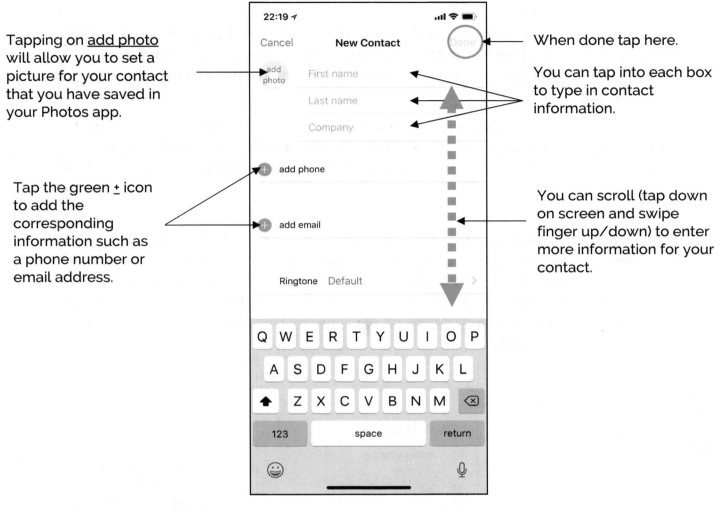

Figure 6.2 – Phone App -> Contacts Tab -> Add New Contact (+)

Browsing Contacts

At any time, you can view your contact list by opening the <u>Phone</u> app and tapping the <u>Contacts</u> tab at the bottom. Now you will see a list of all of your contacts. You can browse through this list by simply tapping down on the <u>screen</u> and scrolling (swiping) up or down.

TIP: To quickly browse through your contacts by their last name. Tap down and scroll on the <u>letters</u> at the far right of your screen.

You can also tap into the <u>search bar</u> at the top of your Contacts screen to search for a contact directly.

To view a contact's information, simply tap on their *name*.

Editing or Deleting a Contact

To edit a contact, tap on their _name_ inside the contact list, and then tap <u>edit</u> at the upper right. Now you can edit all of their information. To delete a certain aspect of their information, simply tap on the <u>red minus sign</u> next to it. When you are done editing a contact tap on <u>Done</u> at the upper right.

To delete a contact, tap on their _name_ inside the contact list. Now tap <u>edit</u> at the upper right. Now scroll down all the way to the bottom of the contact's information, and tap on <u>Delete Contact</u>. A confirmation will appear, where you can tap on <u>Delete Contact</u> again to confirm the deletion.

Favorites List

The favorites list is basically the speed dial of your iPhone. To access you Favorites List, open the <u>Phone</u> app, and then tap the <u>Favorites</u> tab. To add a contact to your Favorites List, first tap the <u>plus sign</u> at the upper left hand corner. Then scroll to the contact you wish to add to your favorites, and then tap on that contact. Now tap on which segment of that contact you want to be on your favorites list, such as <u>text message</u> or <u>phone call</u>. Your contact will now appear in your favorites list. (<u>Figure 6.3</u>)

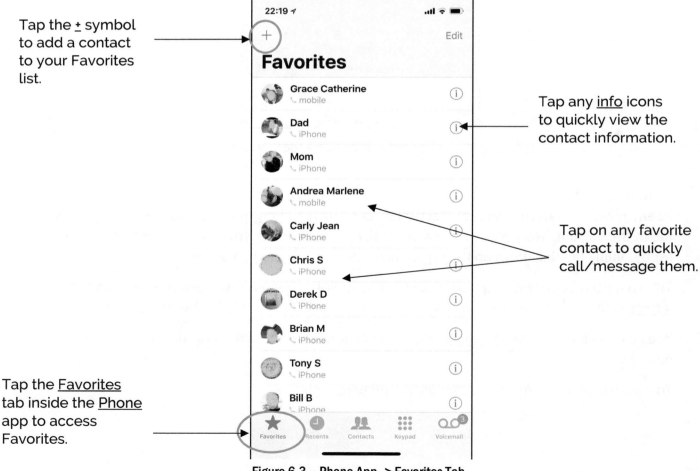

Tap the ± symbol to add a contact to your Favorites list.

Tap any <u>info</u> icons to quickly view the contact information.

Tap on any favorite contact to quickly call/message them.

Tap the <u>Favorites</u> tab inside the <u>Phone</u> app to access Favorites.

<u>Figure 6.3</u> – Phone App -> Favorites Tab

Chapter 7 – Phone Calls

At its heart the iPhone X is a great cell phone. Making and receiving calls is easy and user-friendly, and we will explore that in this chapter.

Making a Call

There are numerous ways to make a call using your iPhone. I will cover some of the most basic ways.

How to Dial a Call (Figure 7.1)

1. Open the Phone app on your home screen
2. Tap the Keypad tab at the bottom
3. Tap in the number you wish to call, one digit at a time
4. Tap the green phone icon to start the call

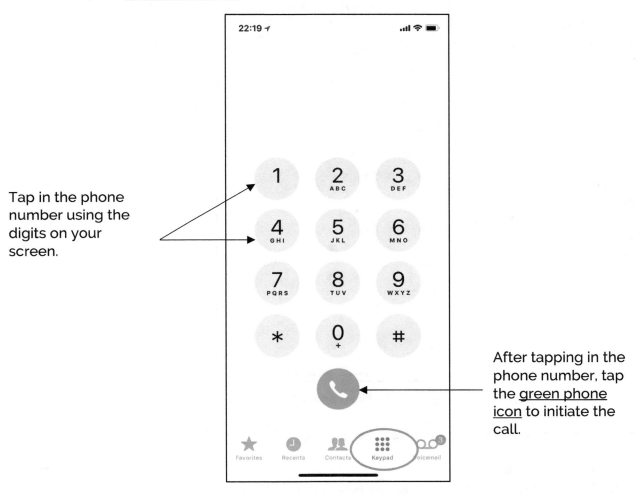

Tap in the phone number using the digits on your screen.

After tapping in the phone number, tap the green phone icon to initiate the call.

Figure 7.1 – Phone App -> Keypad Tab

How to Call Someone on Your Contact List (<u>Figure 7.2</u>)

1. Open the <u>Phone</u> app
2. Tap the <u>Contacts</u> tab
3. Find the contact you wish to call and tap on their *<u>name</u>* to view their contact details.
4. Now, to start the call you can either tap the <u>call icon</u> inside the contact's information or tap on the corresponding <u>phone number</u> within the contact to start the call.

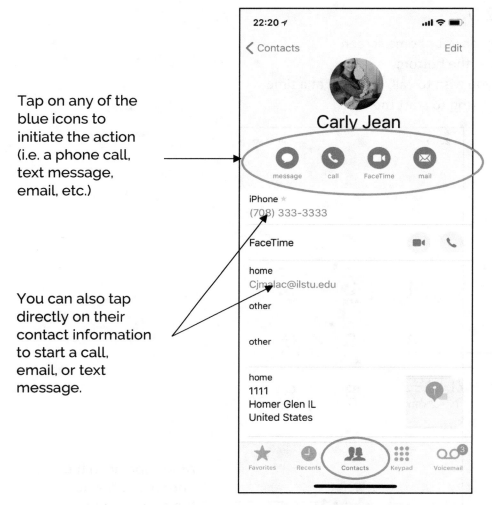

Tap on any of the blue icons to initiate the action (i.e. a phone call, text message, email, etc.)

You can also tap directly on their contact information to start a call, email, or text message.

<u>Figure 7.2</u> – Phone App -> Contacts Tab

How to Call Someone on Your Favorites List (See <u>Figure 6.3</u>)

1. Open the <u>Phone</u> app
2. Tap the <u>Favorites</u> tab
3. Tap on the *<u>contact</u>* in your Favorites List that has a small phone icon underneath their name to immediately start a call with them

Receiving a Call

When you receive a call, a screen will appear (See <u>Figure 7.3</u>). If your iPhone is currently unlocked (i.e. your iPhone screen is on), all you need to do to answer the call is tap on the <u>green phone icon</u>. When you do, you will now be connected to the call.

<u>Figure 7.3</u> – Receiving a Call, iPhone Unlocked

When you receive a call when your iPhone is currently locked, you will see a slider at the bottom of your screen (<u>Figure 7.4</u>). To answer the call, touch down on the <u>green phone icon</u>, and slide it to the right with your finger, releasing at the end. This will connect you to the call.

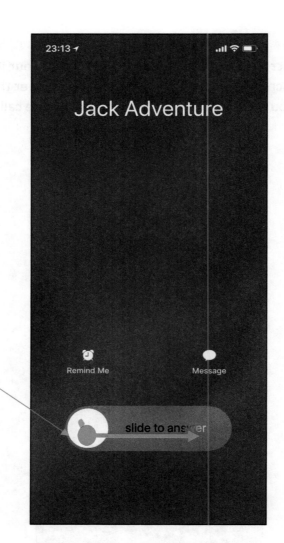

To answer the call, tap down on the green phone icon and slide your finger to the right, releasing at the end

Figure 7.4 – Receiving a Call, iPhone Locked

You can always reject a call by pressing the sleep/wake button on your iPhone when a call is incoming.

Call Functions

During a phone call, you have several functions available to you. Looking at your screen during a call (Figure 7.5), there are several buttons on your screen. They do the following if you tap on them:

- **Mute** – Mutes your side of the call. In other words, the person you are talking to on the phone will not be able to hear anything from your side until you touch mute again and turn mute off.
- **Keypad** – Brings up a touch-tone keypad, in case you need to work through a touch-tone system.

- **Speaker** – Tapping <u>speaker</u> turns on speakerphone. Tap it again to turn it off.
- **Add Call** – Tapping on <u>add call</u> lets you add another person to the call, such as 3-way calling (If part of your service).
- **FaceTime** – FaceTime lets you connect to the person you are talking to over a FaceTime call, which is a video call. This only works when you are talking to someone who has a FaceTime enabled device (more on FaceTime later in this book).
- **Contacts** – Tapping on <u>contacts</u> will bring you to your contact list.

You can also browse through your iPhone while on a call by simply performing a <u>home swipe</u>. To get back to the call screen itself mid-call, tap on the <u>time with a green highlight around it</u> at the top left of your screen.

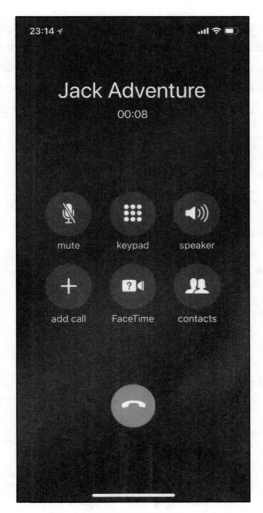

Figure 7.5 – In-Call Options

49

Recent Calls

To access your recent calls list, open the Phone app and touch the Recents tab at the bottom. This screen lists all of your recent calls. For names listed in red, these are missed calls. The times indicated on the right are the times the call took place. To quickly call someone on your recent calls list, simply tap down on their *name* and your iPhone will call them. You can also view more information about a recent call by tapping on the information circle to the right.

Voicemail

(See Figure 7.6)

A common question I get is how to setup voicemail on the iPhone. Unfortunately, there is no easy answer. The reason is because voicemail is completely controlled by your cellular provider. You will have to contact them to set it up.

For some providers, you can setup voicemail from the Phone app. Just open the Phone app, tap on the Voicemail tab at the bottom right, and your voicemails will be listed. You can tap on one to listen to it. You can also set your voicemail greeting by tapping on Greeting at the upper left. Again, this setup will not work for everyone, as it is completely controlled by your cellular provider. Your best bet is to contact them in order to set your voicemail up.

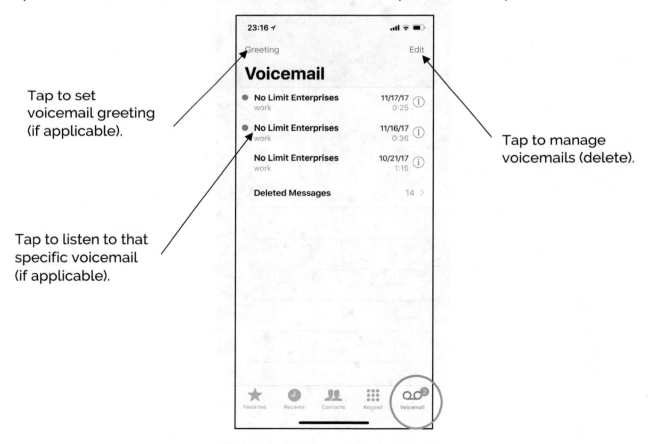

Figure 7.6 – Phone App -> Voicemail Tab

Chapter 8 – Text Messaging

The iPhone is a wonderful tool for text messaging. The ease in which to send and receive text messages is astounding, and the way in which texts are organized is even better. Let's explore the Messages app.

Somewhere on your home screen will be the <u>Messages</u> app. Tap on it to open it up. This is where all of your text messaging will take place, and where all of your text messages will be stored.

Before we dive in, let me explain some texting jargon. When you exchange a text message with someone, it is generally called a conversation or thread. In this book, we will call any text conversation with someone or a group of people a "thread." (<u>Figure 8.1</u>)

Manage messages
(Delete threads)

Start a New Text
Message (Called a
thread)

Unread messages
will have a blue
icon.

These are all your
threads (text
messages), sorted
by person or
group.

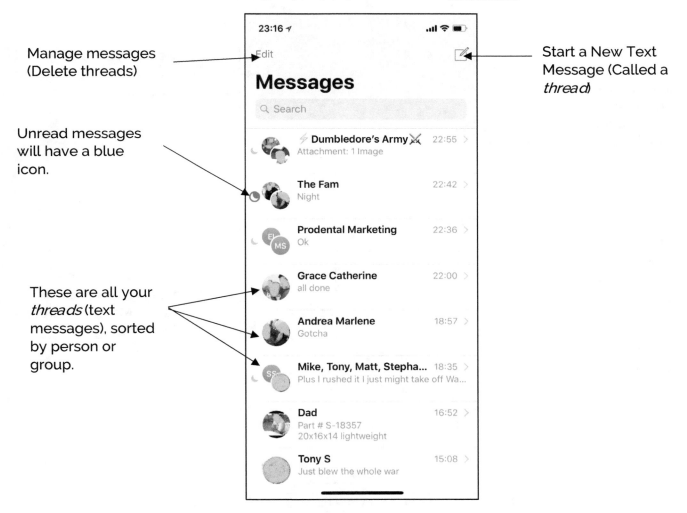

<u>Figure 8.1</u> – Messages App

Sending a Text Message

To send your first text message on your iPhone, first tap the Square and Pencil icon at the upper right of your screen inside the Messages app (Figure 8.1). This will start a new text thread. In the screen that appears, you can type in the name of one your contacts and as you are typing, your iPhone will make suggestions as to whom you are trying to text. Once you see their name, you can tap on their *name* to complete the entry. (Figure 8.2)

Alternatively, you can use your keyboard to type in the exact phone number of someone you want to text.

Next you can tap into the Message bar or iMessage bar near the middle of your screen. Now you can use your keypad to type in a message. Once that message is ready to be sent, tap the blue or green up arrow icon to send the message. (Figure 8.2)

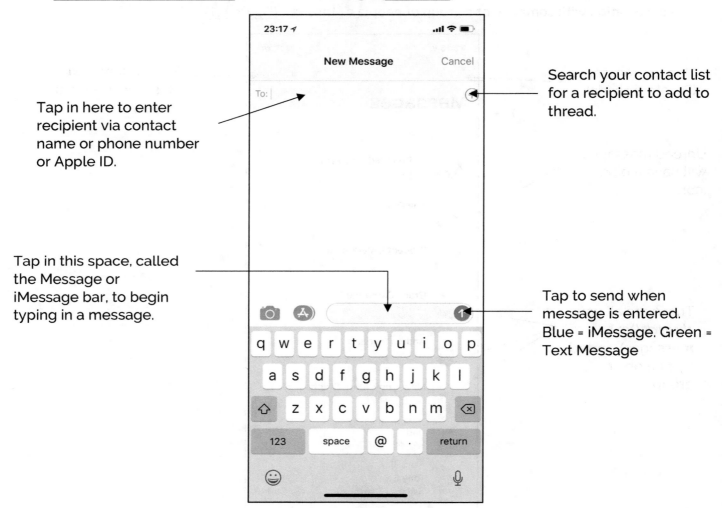

Search your contact list for a recipient to add to thread.

Tap in here to enter recipient via contact name or phone number or Apple ID.

Tap in this space, called the Message or iMessage bar, to begin typing in a message.

Tap to send when message is entered. Blue = iMessage. Green = Text Message

Figure 8.2 – Messages App -> Start New Thread (Pencil & Square Icon)

Receiving a Message

When you receive a message, it will appear in the Messages app (See <u>Figure 8.1</u>). Unread messages will have a blue icon next to the thread. To view the message, open the thread by tapping on the <u>thread's *name*</u>. Your entire text message conversation will now be shown, including any new messages that you have received from this thread. You can scroll up and down through the history of the conversation using your finger. At any time while inside a thread, you can return to the main Messages screen by tapping on the <u>back arrow</u> at the upper left.

Group Threads

Text messaging is not reserved for a single person. You can create group threads where multiple people can exchange text messages. The process is exactly the same:

1. Tap the <u>square and pencil icon</u> at the upper right to create a new message.
2. Enter in the name of a person you want to text of the group. Tap on their <u>*name*</u> once it appears in suggestions.
3. Now type in another name of a person you also want to be a part of the group thread, and tap on their <u>*name*</u> in suggestions when it appears.
4. Add as many people to the thread as you wish.
5. When ready to start conversing, tap into the <u>Message box</u> and begin typing away.

This group thread will now be saved in your Messages app and you can converse in there any time you wish.

iMessage

iMessage is a special term used for text messages exchanged between people who use Apple devices. For instance, you may notice when texting certain people that your messages will appear as the color blue (<u>Figure 8.3</u>). Conversely, you will notice when texting other people that your messages will appear as the color green (<u>Figure 8.4</u>). When they appear blue, that means you are sending iMessages to each other. In other words, you both are using Apple devices such as an iPhone, iPad, or Mac to send your text messages. iMessages (blue) are different because they are not sent as cellular text messages. In other words, they do not count against your text message count if your cellular plan limits them. Also, when you are conversing with someone using iMessage, there are additional options available to you, which we will now cover.

Figure 8.3 – iMessage Example

Figure 8.4 – Regular Text Message

Special Effects

There are a ton of special effects and add-ons you can use with the Messages app (Figure 8.5). For instance, if you want to text message a picture to someone, tap the camera icon to the left of the message bar. Now you can browse through your saved photos on the bottom by swiping left and right, and you can send one of those photos by tapping on a photo and then tapping the send icon. Alternatively, you can take a picture to send by swiping to the left and tapping on Camera. (Figure 8.6)

There are more special effects you can send as well. We will not cover all of them in here, but you can send emoticons by tapping on the smiley face on your keyboard. More effects are available by tapping on the App Store icon. Lastly, you can send voice messages to the thread by using the microphone icon to the right of the message bar, which will only appear when no text is entered into the message bar. Take the time to play around with these features at your leisure to see all you can do with Messages. You will notice that some features can only be used when exchanging iMessages with another Apple user.

This is the *back arrow*, it will always bring you back one screen. (Used in many apps)

See thread details including who is in thread

Name of thread or contact name

Send Photos

Send audio messages

See apps for Messages

Send emoticons

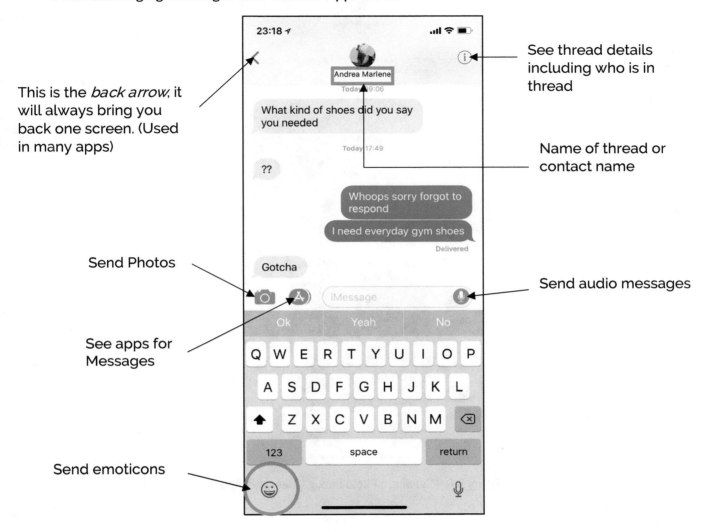

Figure 8.5 – Messages Features including Special Effects

Figure 8.6 – Sending a Photo through Messages App

Managing Messages

An important aspect of messages is managing them. Let's take a look at a couple key management features that are helpful to know.

How to Delete Threads

You can delete an entire thread by following these steps:

1. Open the <u>Messages app</u>.
2. Tap <u>Edit</u> at the upper left
3. Tap each thread you want to delete, until they are check-marked.
4. Tap <u>Delete</u> at the lower right.

How to Delete Individual Texts

1. Open a thread in the <u>Messages app</u>.

2. Find the exact text you want to delete, and tap and hold on it until some options appear. (Figure 8.7)
3. Tap More at the bottom.
4. Now you can tap each individual text within a thread to be check-marked that you want deleted.

5. Tap the trash icon at the lower left to delete the selected texts.

Tap and hold with your finger on an individual text in order to bring up these options. Then tap on More... to select texts for deletion.

You can emote a text by using these icons, similar to "liking" a text. (iMessage only)

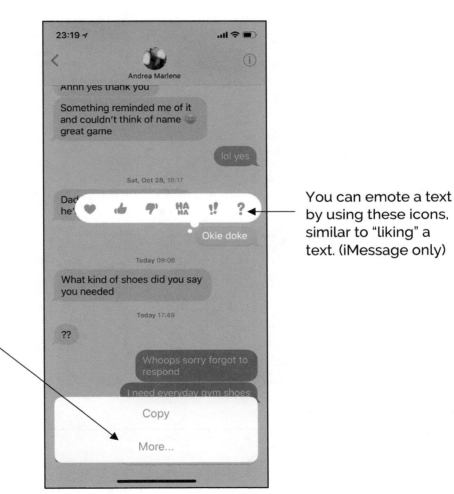

Figure 8.7 – Managing Individual Text Messages

Thread Details

Thread details offer a glimpse into certain information about a thread, such as who is in the thread, and any pictures exchanged in the thread. To access thread information, simply open a thread in Messages, and then tap the information icon at the upper right.

Chapter 9 – Email

Your iPhone has the ability to seamlessly manage your email accounts. It is quite convenient when you can quickly reply to an email in 10 seconds straight from your iPhone. All of your emails and email accounts are located within the Mail app, which we will get to shortly. For now, let's start by adding your existing email account to your iPhone.

Adding an Email Address to iPhone

1. Open the Settings app.
2. Find the Accounts & Passwords tab, and tap on it.
3. Tap Add Account.
4. Look through the list of pre-populated email accounts such as AOL, Google (Gmail), and Yahoo. If you have one of these accounts, tap on it. If your account does not appear, tap on Other. (Figure 9.1)
5. If you tapped on a pre-populated account, follow the instructions that appear on your screen. Most likely you will just have to enter in your email address and password. If you tapped on Other, tap on Add Mail Account. (Figure 9.2)

Figure 9.1 – Settings -> Accounts & Passwords ->
Add Account

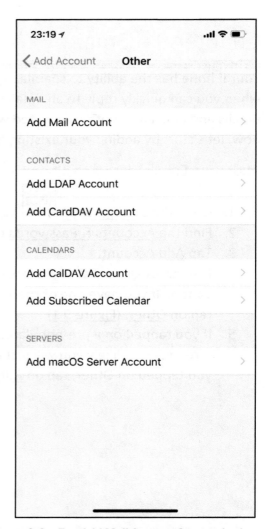

Figure 9.2 – Tap Add Mail Account for standard
email addresses, if you chose Other.

6. Now type in the requested information. When done, tap Next at the upper right.
(Figure 9.3)

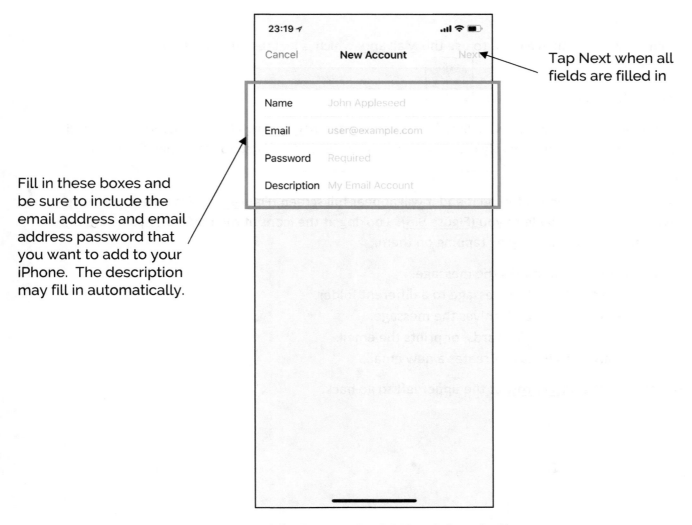

Fill in these boxes and be sure to include the email address and email address password that you want to add to your iPhone. The description may fill in automatically.

Tap Next when all fields are filled in

Figure 9.3 – Enter your Email Address Information Here

7. Your iPhone will attempt to add your email given the information you provided. For most email accounts, this should work with no issues. However, for less common email accounts, your iPhone may require more information such as incoming and outgoing server addresses. If this is the case for you, you will have to contact your email provider in order to get this information.

8. Continue following the instructions that appear on your screen until you have reached the end. You can continue by tapping on Next/Done at the upper right.

Once your email account has successfully been added, you are ready to use email on your iPhone. You can add multiple email addresses to your iPhone if you wish.

Checking Your Email

To check your email, we need to use the <u>Mail app</u>, which is located somewhere on your home

screen. Tap on it to open it up.

Your email accounts will now be listed. You can use the <u>back arrow</u> at the upper left to view all of your email accounts. Simply tap on an account to view your email for that specific account.

Viewing Email

To view an email, simply tap on it and it will appear full screen (<u>Figure 9.4</u>). From here you have several options available to you (<u>Figure 9.5</u>). Looking at the icons at the bottom of your screen, they perform the following by tapping on them.

- **Flag** – Flags or marks the message.
- **Folder** – Moves the message to a different folder.
- **Trash** – Deletes or Archives the message.
- **Arrow** – Replies, forwards, or prints the email.
- **Square with Pencil** – Creates a new email.

You can use the <u>back arrow</u> at the upper left to go back.

Tap to go back

Manage email in bulk

The blue dot icon means unread mail

Tap on any email message to read it

Create new email message

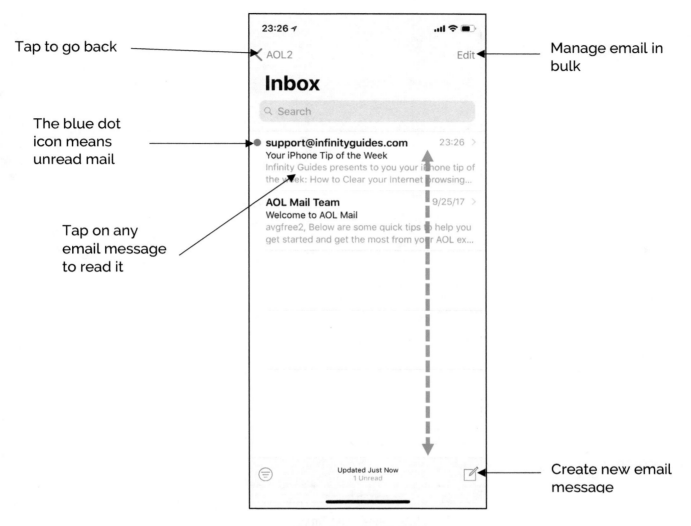

Figure 9.4 – The Mail App

Tap to go back

Tap to quickly view next or previous email

Email controls

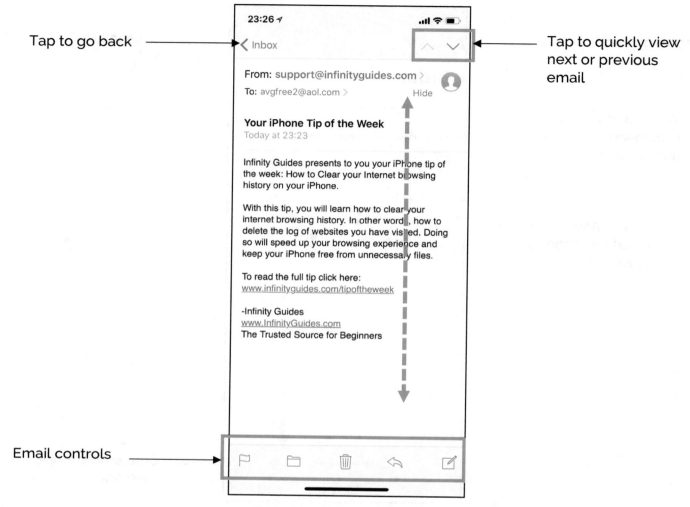

Figure 9.5 – Viewing an Email

Sending Email

At any time from the Mail app, you can create a new email by tapping the <u>square and pencil icon</u> at the lower right. This will bring up a new message box. Full steps are (<u>Figure 9.6</u>):

1. Tap the <u>square and pencil icon</u> at the lower right inside the <u>Mail app</u>,
2. Enter in the email address or contact name of the intended recipient. Use the <u>+</u> icon to search through your contacts.
3. Tap into the <u>subject line</u>, and type in the subject.
4. Tap into the <u>message body</u>, and type in the email message.
5. Tap <u>Send</u> at the upper right, or tap <u>Cancel</u> at the upper left to cancel the message.

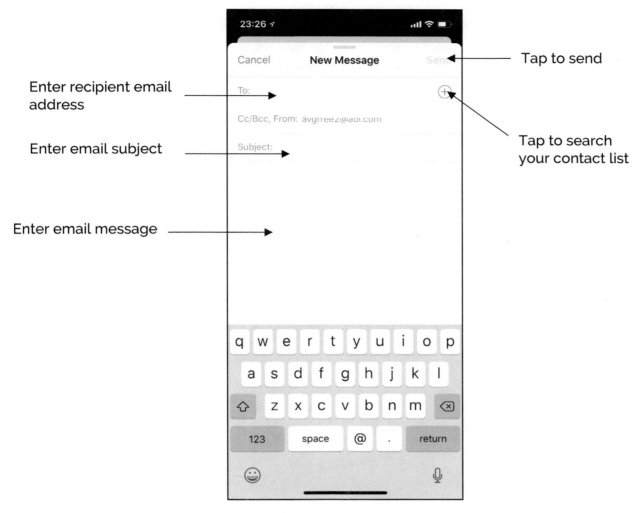

Enter recipient email address

Enter email subject

Enter email message

Tap to send

Tap to search your contact list

Figure 9.6 – Composing New Email

Managing Email
To Delete Email(s)

1. Tap an email account or folder to view all emails in that category.
2. Tap Edit at the upper right.
3. Tap each email you want to manage.
4. Use the text at the bottom to manage. Tap on Mark, Move, or Archive/Delete/Trash.
5. To cancel, tap Cancel at the upper right.

TIP: You can quickly delete an individual email by tapping on its preview, then swiping to the left.

Chapter 10 – Web Browsing

One of the signature features of the iPhone is the ease in which you can surf the web. To access the Internet, open the <u>Safari app</u> on your home screen.   Safari is where you will be doing all of

your internet browsing, unless you download a different app for internet browsing.

Visiting Web Pages

Safari is Apple's web browser optimized for the iPhone. In Safari, you can visit web pages as well as search the web using your favorite search engine. By default, Google is set as the current search engine. To search the web, tap in the box at the top of the screen and type in your search parameters, then touch <u>GO</u> on your keypad. If you want to visit a webpage directly, touch in the same box and type in the web address, followed by tapping <u>GO</u>. (<u>Figure 10.1</u>)

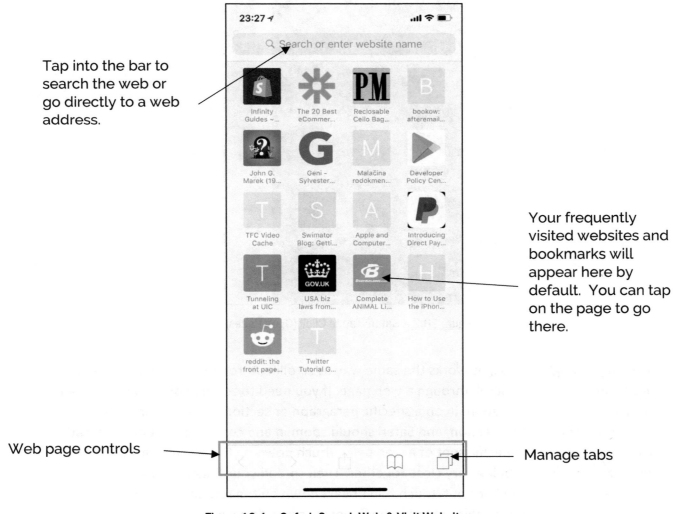

Figure 10.1 – Safari: Search Web & Visit Websites

Tap into the bar to search the web or go directly to a web address.

Your frequently visited websites and bookmarks will appear here by default. You can tap on the page to go there.

Web page controls

Manage tabs

Navigating the web on the iPhone is a unique experience and takes some practice. Think of your finger as the mouse pointer, and touching down on the screen as a click. To open a link, simply touch down on it with your finger (Figure 10.2).

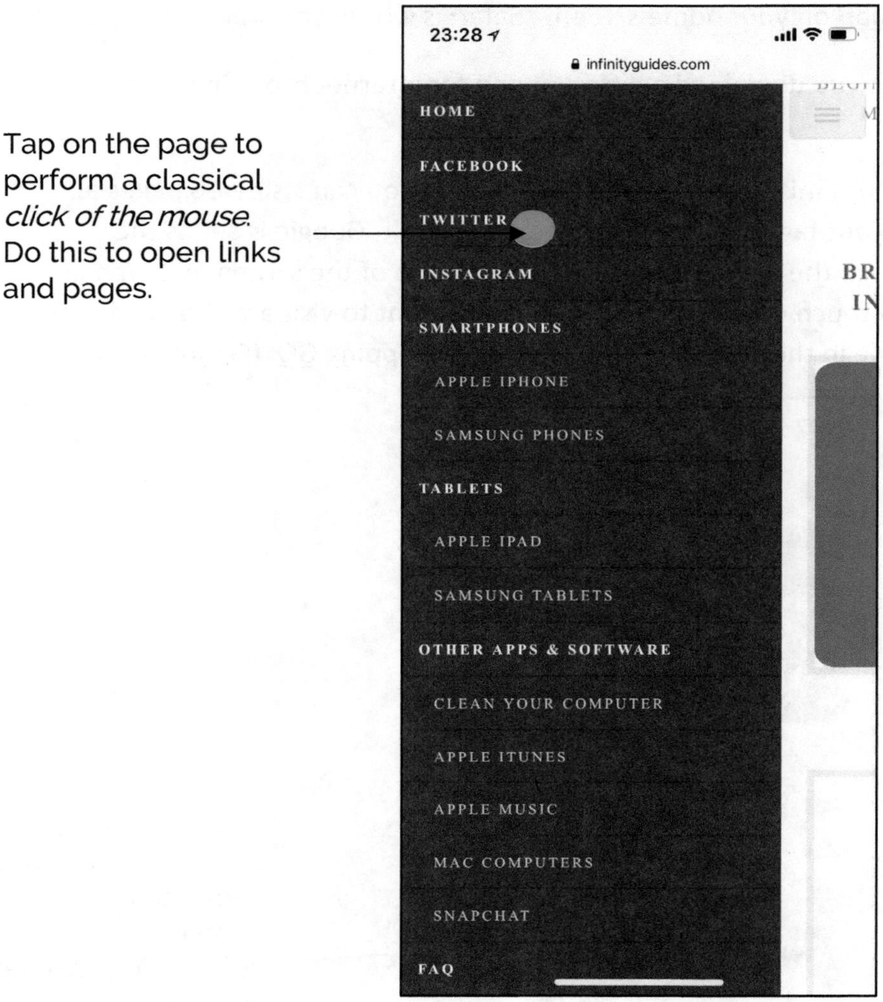

Tap on the page to perform a classical *click of the mouse*. Do this to open links and pages.

Figure 10.2 – Safari: Tap to Click (Open Pages)

Scrolling through web pages works the same way as scrolling through apps. Simply drag your finger up and down to scroll through a web page. If you need to zoom in on something, there are 2 different ways. To zoom in on a specific paragraph or section of a web page, double tap with your finger on that section, and Safari should zoom in and center the screen automatically. To zoom in on a very specific spot of a web page, touch down on the screen with two fingers, with your two fingers being very close together. Then spread your fingers apart while remaining on the screen. To zoom back out, touch two fingers down on the screen again, only this time have the fingers start far apart, and pull the fingers in towards each other. (Figures 10.3 & 10.4)

Double tap with ONE finger on the screen, and Safari will attempt to zoom to a perfect fit, centered on the section you double-tapped.

To zoom out: touch down on the screen with TWO fingers, with the fingers spaced far apart. Then drag the two fingers together while remaining on the screen. Release when the fingers come together, or when you are satisfied with the zoom.

Figure 10.3 – Safari: Double Tap to Fit to Screen & Zoom Out

To zoom in: touch down on the screen with TWO fingers, with the two fingers being close together. Then drag the fingers away from each other in opposite directions while remaining on the screen, releasing at the end or when you are satisfied with the zoom.

Figure 10.4 – Safari: Zooming In

You can still do all the basic functions of your normal web browser with Safari. To browse through basic web functions, touch down at the bottom of your screen. This will bring up Safari navigation options also known as web page controls (See Figure 10.1). From here, you can browse as you please. For instance, to go back to the previous page, touch the left arrow. To go forward a page, touch the right arrow. To manage a webpage, touch the rectangle with the arrow in it. From here you have the options to mail the webpage to a friend, message the page to a friend, tweet the page on your Twitter, share the page to Facebook, add a shortcut to the web page to your home screen, print the web page, bookmark the page, or to add the page to your reading list (Figure 10.5). If you choose to bookmark the page, the page will appear in your bookmarks screen which can be accessed by touching the book icon at the bottom of Safari.

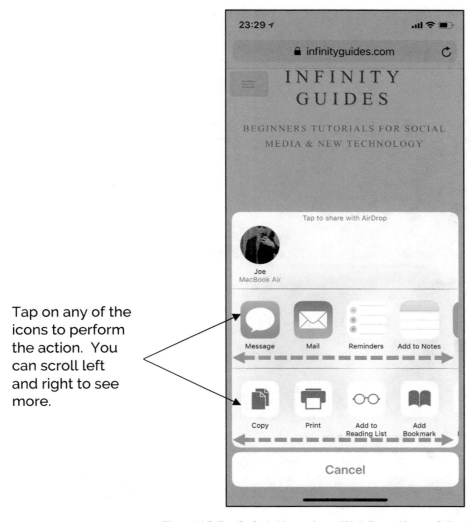

Tap on any of the icons to perform the action. You can scroll left and right to see more.

Figure 10.5 – Safari: Managing a Web Page (Arrow & Rectangle Icon)

Finally, you can browse in multiple tabs in Safari, by tapping the two windows icon at the bottom right of the Safari screen and then touching the plus icon (Figure 10.6). Tabs are just additional windows where you can browse the internet; an excellent tool for multi-taskers. To browse between tabs, touch the windows icon again. To truly master browsing through Safari, just take some time playing with it, and you will get it in no time.

Tap the x or swipe the page off the screen to close a tab.

Creates a new tab

Switches you to private browsing (does not save cookies or history)

Exits tab manager and returns to Safari main screen

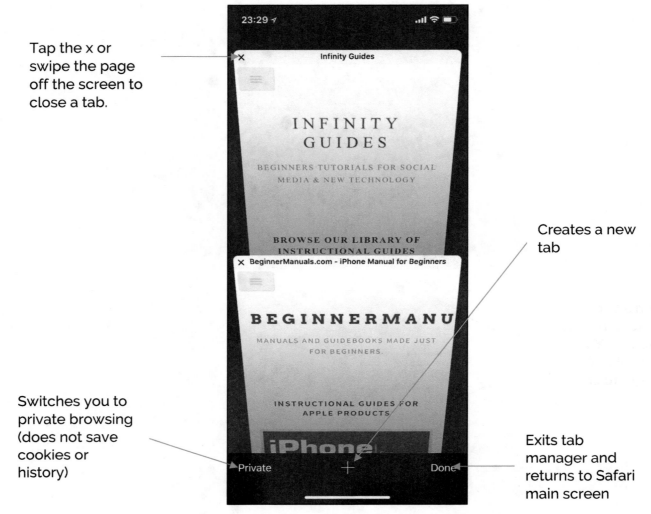

Figure 10.6 – Safari: Using Tabs (Windows Icon-See Figure 10.1)

Chapter 11 – Using your Camera

The iPhone X is equipped with a powerful camera with flash capability. To access the camera, touch the Camera app on your home screen. ◻ There are a few options available inside the Camera app. On the top of your screen from left to right is: (Figure 11.1)

- **Lightning:** This turns flash to on, off, or auto.
- **Concentric Circles –** This enables or disables LIVE photos, which capture a second or so before and after the photo, thus creating what is called a "live photo."
- **Timer –** Tapping on this icon allows you to set a timer for the capture sequence.
- **Color Palette –** Allows you to choose a filter to capture the photo.

Taking a Photo

To take a photo, aim your camera and then tap on the large circle icon at the bottom of your screen. You can also turn your iPhone to the side to take a picture in landscape mode. Furthermore, you can use the selfie lens, which faces directly back at you, by tapping on the camera icon at the lower right of your screen. (Figure 11.1)

Camera Options

You can switch between different camera modes by tapping down anywhere on your screen while in the Camera app, and then swiping to the left and the right. You will notice different text become highlighted right above the capture icon to indicate which mode you are in. (Figure 11.1)

The different modes are:

- **Photo –** Standard photo mode
- **Video –** Records a video
- **Slo-Mo –** Records a video in slow motion
- **Time-Lapse –** Records a time lapse sequence
- **Portrait –** Ideal for taking professional portrait photos (Advanced)
- **Square –** Captures a photo in a perfect square
- **Pano –** Captures a panoramic photo

You can preview the photo you have just taken by tapping on the preview of the photo at the lower left of your screen. To get back to the camera from this preview, tap Camera at the upper left.

Camera Options

Tap on screen
and swipe left or
right to switch
between
camera modes

Camera Modes
(highlighted is
current mode)

Recent Photo Preview
(tap to load)

Switch between regular and
selfie lens

Capture Icon (tap to take
picture or record video)

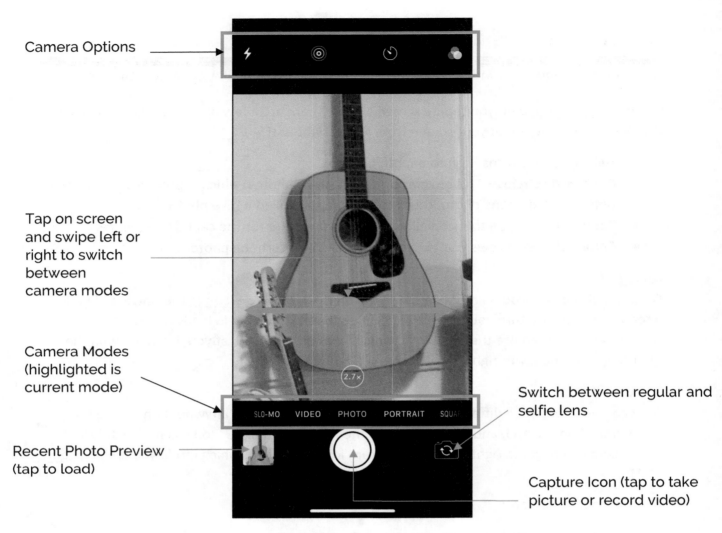

Figure 11.1 – Camera App

74

Chapter 12 – Photos & Videos

Now that we know how to take photos and record videos, it is time to learn what we can do with these. All of your photos and videos can be found in the <u>Photos app</u>, which is located

somewhere on your home screen.

Photos App Layout

When you first open the Photos app, it may look like there is a lot going on (<u>Figure 12.1</u>). Let's take a look at navigating this expansive app. At the bottom will be your browsing tabs. The <u>Photos</u> tab will show you all of your photos. You can browse through these by scrolling up and down. <u>Memories</u> will show a specific memory, which are automatically generated by your device. An example of a memory is a photo you took one year ago from today. The <u>Shared</u> tab will bring you to shared photo streams, which we will discuss a little later on. Lastly, the <u>Albums</u> tab will bring you to all of your albums. At any time you can view a photo in full-screen by tapping on it.

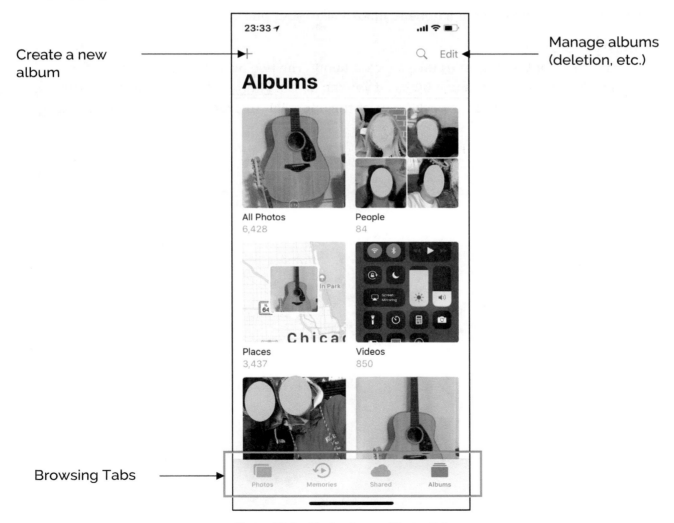

Create a new album

Manage albums (deletion, etc.)

Browsing Tabs

Figure 12.1 – Photos App -> Albums Tab

iCloud Photo Library

Before we explore the Photos app more thoroughly, it is important to discuss what is called iCloud Photo Library. Photos take up a lot of storage space on your iPhone. Videos take up even more space. Your iPhone is equipped with what is called iCloud Photo Library, which basically stores your photos off of your iPhone and in the cloud, which is just a hard drive that is stored on the internet. iCloud Photo Library may be a great option for you if you plan on keeping a lot of photos on your iPhone. In order to use iCloud Photo Library your Apple ID must be set up, which we covered at the beginning of this book. If you want to enable iCloud Photo Library, follow these steps:

1. Open Settings app
2. Find Photos and tap it
3. At the top, tap to the right of iCloud Photo Library to enable it (green will show).
4. Choose whether you want to Optimize iPhone Storage or Download and Keep Originals. (Optimize storage will save you space)
5. You will now be using iCloud Photo Library.

It may take some time for your iPhone to upload all of your photos into the cloud.

Browsing Photos

Back inside the Photos app, let us take a look around. The two best tabs at the bottom for browsing photos are Photos and Albums. If you tap on the Photos tab, all of your photos will be sorted by date. You can use the back arrow at the upper left to expand your view. (Figure 12.2)

Back arrow
(expand view)

Manage photos
(delete and
select multiple)

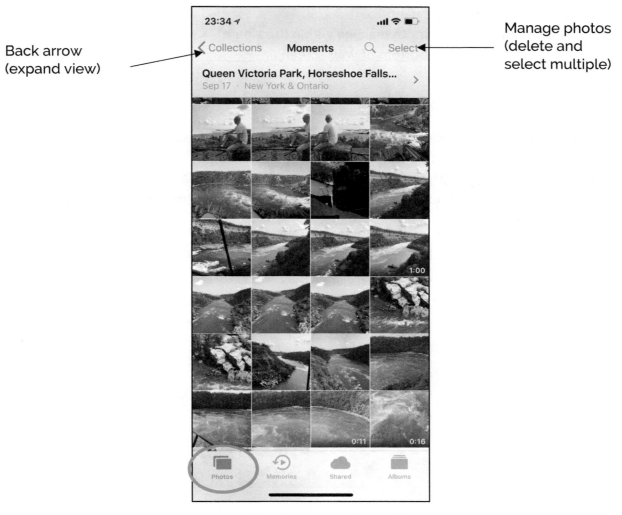

Figure 12.2 – Photos App -> Photos Tab

Tapping on the <u>Albums</u> tab, you can browse all of your photos as thumbnails by tapping on the <u>All Photos</u> album. To view a photo, tap on <u>*it*</u> to bring it full screen. Use the <u>back arrow</u> at the upper left to go back. (<u>Figure 12.1</u>)

To Create an Album
(See <u>Figure 12.1</u>)

1. Tap the <u>Albums</u> tab in Photos
2. Tap the <u>plus sign</u> at the upper left (If you do not see this, tap on <u>Albums</u> again)
3. Type in your new album name and tap <u>Save</u>
4. Scroll through your photos and tap each one you want to add to the album
5. Once you have selected all the photos you want, tap <u>Done</u> at the upper right

To Edit an Album

1. Open the album you want to edit in the <u>Albums</u> tab by tapping on <u>*it*</u>

2. Tap <u>Select</u> at the upper right
3. Tap each photo you want to manage within the album
4. Use the options at the bottom of your screen

Sharing a Photo

To share a photo, follow these steps (<u>Figure 12.3</u>):

1. Find the photo you want to share, and tap on <u>it</u> to bring it full screen. (Note: You can select multiple photos by tapping <u>Select</u> at the upper right).
2. Tap the <u>rectangle with the arrow inside it</u> at the lower left.
3. Tap on the corresponding option as to how you want to share it (<u>Message</u>, <u>Mail</u>, <u>Facebook</u>, etc.).

<u>Figure 12.3</u> – Sharing a Photo

Deleting Photos

1. Tap on the *photo* you want to delete (or select multiple by tapping on <u>Select</u> at the

> upper right).
2. Tap the trash icon at the lower right to delete.

Editing a Photo
(See Figure 12.4)

Remove red-eye

Auto-adjust

Editing options

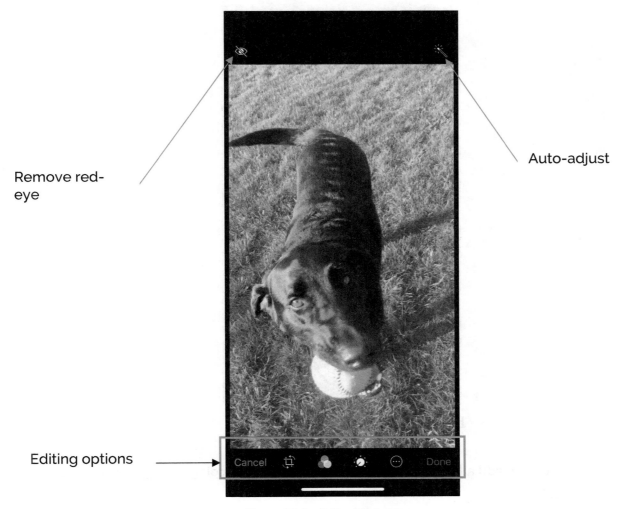

Figure 12.4 – Edit a Photo Menu

You can edit a photo in multiple ways, here's how:

1. Tap on the *photo* you want to edit to bring it full screen.
2. Tap Edit at the upper right.
   a. You can crop the photo by tapping on the crop icon at the bottom of your screen. Then use your fingers and drag the border to determine the crop. You can rotate the photo using the block and arrow icon. Tap the crop icon again to return. (Figure 12.5)

Tap and drag
the corners of
the picture until
your desired
crop has been
reached. Tap
<u>Done</u> when
finished.

Rotate photo

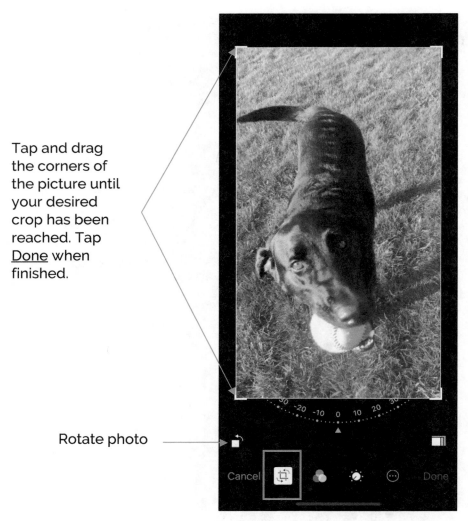

<u>Figure 12.5</u> – Cropping and Rotating a Photo

b. You can add a filter to the photo by tapping on the <u>color palette icon</u>. (<u>Figure 12.6</u>)

80

Tap on a filter
choice to add
the filter. Tap
Done when
finished.

Figure 12.6 – Adding a Filter to a Photo

c. You can alter lighting and shading of the photo by tapping on the <u>sun icon</u>.
    (<u>Figure 12.7</u>)

Drag the slider
to the desired
setting

Figure 12.7 – Adjusting Lighting and Color Settings of a Photo

d. You can markup the photo by tapping on <u>the 3 dots within the circle icon</u>. Tap <u>Markup</u>. (<u>Figure 12.8</u>)

Using markup, you can write directly on the picture by placing your finger on the photo and drawing with your finger.

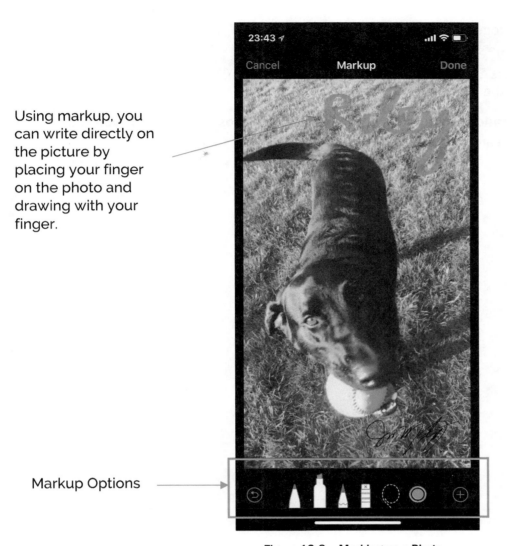

Markup Options

Figure 12.8 – Marking up a Photo

> e. Tap Done when finished with Markup.
3. Tap <u>Done</u> on your screen when you are done editing the photo.

Photo Streams

A Photo Stream allows you to share a select set of photos with your friends who use Apple devices. To create a new Photo Stream:

1. Tap the <u>Shared tab</u> at the bottom inside the <u>Photos app</u>.
2. Tap the <u>back arrow</u> at the upper left.
3. Tap the <u>+</u> sign at the upper left.
4. Name your shared album.
5. Tap <u>Next</u> in the box.
6. Enter in the contacts you want to share the album with one at a time. NOTE: They must be Apple Device users.

7. Tap <u>Create</u> when done.
8. A notification will be sent to all of the contacts you entered.
9. Now tap on the *new shared album* you created. From here, you can add photos to the album by tapping the <u>plus symbol</u>.

When you receive an invitation to a shared Photo Stream, the Photos app will notify you and you can join the Photo Stream from the <u>Shared</u> tab.

Chapter 13 – iPhone X Security

Security is a very important feature of the iPhone X. As I emphasized early on in this book, it is extremely important to remember your lock screen passcode. If you forget this passcode, it is extremely tedious to get into your device. Doing so actually involves resetting your device to factory conditions, which will delete all of your data. In this chapter I will show you how to change the security settings of your iPhone, including your lock screen passcode. Also, please remember your lock screen passcode is different from your Apple ID password.

Settings

All of your security settings can be viewed and changed inside the <u>Settings app</u>. We have used the Settings app several times already in this guide so you should have no problem by now finding and opening the Settings app. As a reminder, the Settings app can be found on your home screen, and you can open it by tapping on it with your finger.

Here inside the Settings app is where you can alter all the under the hood aspects of your iPhone. We will explore more in Settings later on, but for now, locate <u>Face ID & Passcode</u> in and tap on it.

To get into <u>Face ID & Passcode</u> you will have to enter your current Passcode. You may have set one up when you first setup your iPhone X. Enter in your passcode and you will be brought to a new screen.

In this new screen you can alter many settings. You can choose whether to use Face ID, change your passcode, or even turn your passcode off. Let's explore these. (<u>Figure 13.1</u>)

Choose what
you want to be
able to use Face
ID for

Completely
resets Face ID.
You will have to
scan your face
again.

Change or turn
your passcode
off

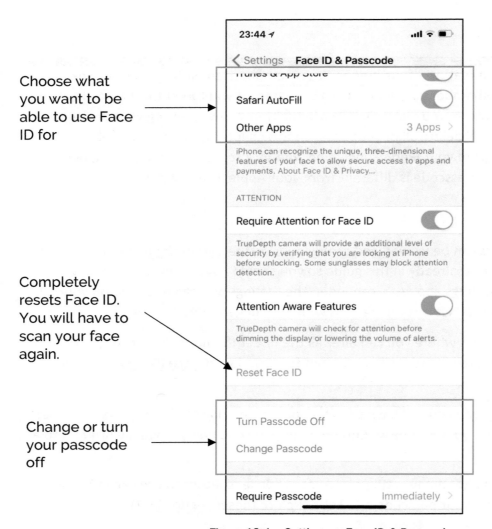

Figure 13.1 – Settings -> Face ID & Passcode

Face ID

Face ID is one of the signature features of the iPhone X, and it allows you to use your face to unlock your iPhone. It also allows you to use your facial scan for some other security measures, such as confirming a purchase of an app or song. When you first set up your iPhone X, you were asked to setup Face ID. To use Face ID to unlock your iPhone from the sleep state, simply tap the screen to turn it on, and then perform a home swipe and look at your iPhone until the unlock symbol appears at the top (Figures 13.2.1 and 13.2.2).

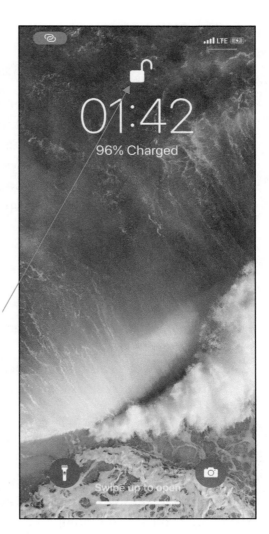

Indicates the iPhone is locked, and you must use Face ID or your Passcode to unlock it.

Indicates the iPhone is unlocked (after Face ID), and you can perform a home swipe to access your home screen.

Figure 13.2.1 – Lock Screen, iPhone Locked

Figure 13.2.1 – Lock Screen, iPhone Unlocked (After using Face ID)

Once your iPhone is unlocked, if you already performed a home swipe you will be brought to your home screen. If you have not yet performed a home swipe, you can do so to access your home screen. If Face ID fails, you will be asked to enter your Passcode to unlock your iPhone. You will also need your passcode whenever you restart your iPhone.

*Please note that when you use Face ID, you will still be required to have a passcode set as well. This passcode will be required whenever your restart your iPhone or if Face ID fails.

Setting up Face ID

As I briefly mentioned, when you first set up your iPhone X, you were prompted to setup Face ID. If you have not done so or want to reset Face ID because it is not working to your standards, here is how:

How to Setup Face ID

1. Open the <u>Settings</u> app.
2. Tap <u>Face ID & Passcode</u>.
3. If prompted, enter your current Passcode.
4. Tap <u>Reset Face ID</u> if you want to reset it or <u>Set Up Face ID</u> if you want to turn it on.
5. Follow the instructions on your screen to setup Face ID. You will have to look directly at your iPhone's screen and move your head around in a circle twice.
6. Once complete, Face ID will be setup and all ready to use.
7. If you previously did not set up a passcode, you will be required to do so when using Face ID.

How to Turn Face ID Off

1. Open the <u>Settings</u> app.
2. Tap <u>Face ID & Passcode</u>.
3. If prompted, enter your current Passcode.
4. Tap <u>Reset Face ID</u> to turn Face ID off.
5. Alternatively, you can turn Face ID off for certain functions by using the options at the top of your screen and disabling them (See <u>Figure 13.1</u>).

Troubleshooting Face ID

On occasion you may experience some issues with Face ID not working properly, especially when you first start using it. Although this may become frustrating, it is important to note that each time you use Face ID, your iPhone is learning your face and becoming better at recognizing it. Therefore, you can expect to see Face ID improve over time and experience less failures. However, if you are continuing to experience numerous Face ID failures, try resetting Face ID in Settings.

Possible Reasons Face ID May Fail

- Environment is too dark
- Large obstruction over your face such as sunglasses or a hat
- Looking at your Face ID at a new angle, such as when laying down or when your iPhone is flat on a surface

Passcode

Your passcode is a password that allows you to unlock your iPhone X in addition to Face ID. Unlocking is simply the process of getting into your iPhone to use it. A passcode is required in order to use Face ID, and if you choose not to use Face ID, I recommend setting up a passcode anyways. Otherwise, anyone will be able to get into your iPhone and have access to all of your data should they ever get possession of your iPhone and your iPhone does not have Face ID or a passcode set up.

Setting or Changing a Passcode

1. Within <u>Face ID & Passcode</u> in <u>Settings</u>, tap <u>Turn Passcode On</u> or <u>Change Passcode</u>.
2. If you are changing your passcode, you will be required to enter your current passcode. Do so.
3. Now you can enter a new passcode. The standard is a 6-digit passcode, however many people prefer to use 4-digits or even an alphanumeric code. I personally prefer to use 4-digits. To change the type of passcode you want, tap on <u>Passcode Options</u>.
4. Now you can choose which type of passcode you want.
5. Tap in your new passcode.
6. Confirm the new passcode by tapping it in again.
7. Follow the instructions on the screen to finish entering your passcode.
8. When complete your passcode will be changed or set.

To Turn Off Passcode Completely

1. Within <u>Face ID & Passcode</u> in <u>Settings</u>, tap <u>Turn Passcode Off</u>.
2. Tap <u>Turn Off</u>
3. Enter in your current passcode.
4. Please note that if you turn passcode off completely, you will not be able to use Face ID.

As a final reminder, you can use your passcode to unlock your iPhone from the sleep state if your Face ID fails.

TIP: REMEMBER YOUR PASSCODE! Write it down somewhere safe and do not lose it! If you forget your passcode it is a very big headache to get back into your iPhone and you risk possibly losing all of your data. If there is one tip I cannot emphasize enough it is this: do not forget your passcode.

Chapter 14 – Personal Settings

There are a number of personal settings you can set on your iPhone, and in this chapter we will explore some of them.

Setting your Wallpapers

Your wallpaper is just your background image on your iPhone. You have 2 different wallpapers: your home screen and your lock screen. To set your wallpapers follow these steps:

1. Open the Settings app.
2. Tap Wallpaper.
3. Tap Choose a New Wallpaper.
4. You can now choose between Dynamic, Stills, Live, and one of your Photos. Dynamic wallpapers are animated on your screen. Stills are not animated and are pre-loaded on the iPhone. Live wallpapers are animated when you press down on them using 3D Touch (More on 3D touch later). You can select one of your current photos by browsing through your albums at the bottom. (Figure 14.1)

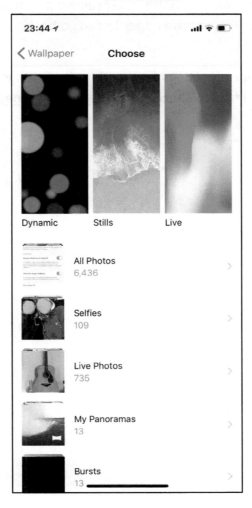

Figure 14.1 – Settings -> Wallpaper -> Choose a New Wallpaper

5. Select a photo to use as your wallpaper.
6. Select if you want the wallpaper to be <u>Perspective</u> or <u>Still</u>. Perspective adjusts to which angle you are viewing your iPhone at.
7. Tap <u>Set</u>.
8. Choose to set the wallpaper as your <u>Lock Screen</u>, <u>Home Screen</u>, or <u>Both</u>.

Ringtones & Vibrations

You can customize your ringtones and vibration settings in multiple ways. Here's how:

1. Open <u>Settings</u>.
2. Tap <u>Sounds & Haptics</u>.
3. Here you can adjust various sound settings such as ringer volume. To change a specific tone, such as your ringtone, tap on <u>Ringtone</u>.
4. Now you can tap on each tone to preview it on your iPhone. Once you have found the tone you want, use the <u>back arrow</u> at the upper left to go back.
5. You can adjust the vibration pattern of a tone by selecting the tone type (i.e. <u>Ringtone</u>) then tapping <u>Vibration</u> at the top.
6. Now you can tap each vibration pattern to preview it. Once you have found the pattern you like, use the <u>back arrow</u> at the upper left to go back.

Other Settings

You can adjust other personalized settings from within the Settings app. We will not cover them here in this text but feel free to check out <u>Display & Brightness</u> and <u>Battery</u> settings.

Chapter 15 – The Home Screen

By now, you should be pretty familiar with your home screen. As a reminder, your home screen is the main screen of your iPhone that shows all of your apps. You have multiple home screens, and you can switch between them by tapping down with your finger on the screen and swiping left or right. To get back to the main home screen at any time, simply perform the home swipe. (See Figure 4.1 in Chapter 4 as a reminder)

3D Touch

3D Touch, also commonly referred to as Force Touch, is a feature on your iPhone that allows you to perform different functions while pressing down on your screen with different pressure. This is a fairly new feature of the iPhone and has not become widely used yet. Let's try it out and see what we can do.

3D Touch Example 1: On Apps

1. On your home screen, locate the Messages app.
2. Instead of tapping and releasing the Messages app to open it, tap down on it with some additional finger pressure and hold down for a moment (This is called *3D Touch*).
3. You should feel your iPhone vibrate and a small popup window will appear (Figure 15.1).
4. Now, you can release your finger.
5. The screen that appears shows some recent contacts you have exchanged text messages with. You can tap on any of their names to immediately be brought to that particular Messages thread.
6. You can return to your home screen by performing a home swipe.

You can use 3D Touch on just about every app. Try it out on other apps to see which shortcuts appear.

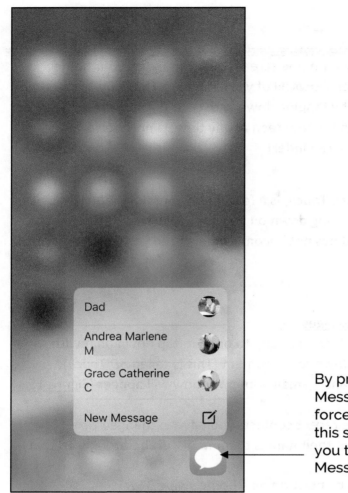

By pressing down on the Messages app with a little extra force and holding for a moment, this screen will appear, allowing you to quickly go to a recent Messages thread

Figure 15.1 – 3D Touch on the Messages App

3D Touch Example 2: Photos

1. Open the Photos App.
2. Find a photo you want to quickly examine and instead of tapping on it normally, tap down and press with some pressure (3D Touch), and do not release your finger.
3. This will open a quick preview of the photo. If the photo is a Live Photo, it will play through it. Furthermore, if you selected a video, this will play the video.
4. If you release your finger, the preview will disappear. If while holding down, you slide your finger upwards and release, some options will appear. (Figure 15.2)
5. The shortcuts that appear are pretty useful, especially the copy function.

This is another example of how you can use 3D Touch. Try using it within other apps as well.

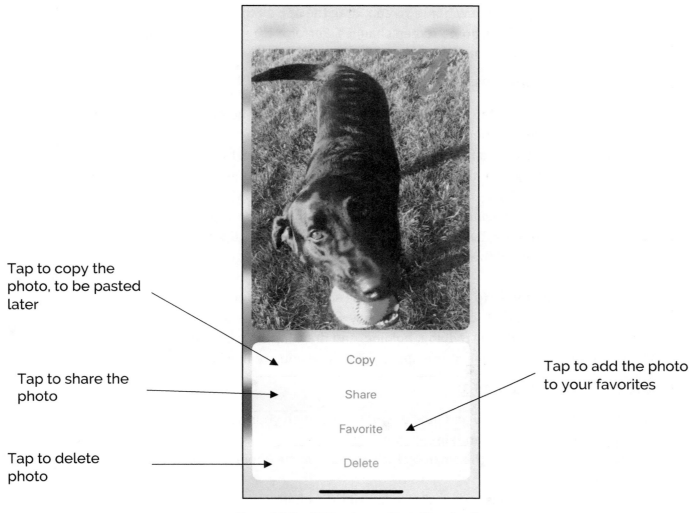

Tap to copy the
photo, to be pasted
later

Tap to share the
photo

Tap to delete
photo

Tap to add the photo
to your favorites

Copy

Share

Favorite

Delete

Figure 15.2 – 3D Touch on a Photo Thumbnail

Organizing Apps

Now that we learned how to use 3D Touch, let us move on to how we can organize our home
screens. On your home screens, you can choose where you want each app to appear. You can
also group multiple apps together in a folder.

Moving Apps Around on your Home Screen (Figure 15.3)

1. With light pressure, tap and hold down on an app on your home screen until all of your
   apps start to shake, then release your finger. (If a 3D Touch popup appears, you will
   have to try again, with lighter finger pressure).
2. Now that your apps are shaking, you are allowed to move the apps around.
3. To move an app, tap down on a shaking app and do not let go. You can now drag your
   finger anywhere on the screen to move the app to a new location.

4. To move the app to a new home screen, drag the app all the way to the right or left and hold it there, until the home screen changes. Perform a <u>home swipe</u> or tap <u>Done</u> at the upper right when you are finished.

Grouping Apps into Folders (<u>Figure 15.3</u>)

1. With light pressure, tap and hold down on an app on your home screen until all of your apps start to shake, then release your finger.
2. If you want to create a folder of apps on your home screen, tap down on an app and hold it, then drag it on top of another app that you want to group with the app you are holding.
3. You will see a small window appear behind your app. Release your finger.
4. A new folder has just been created. Your iPhone will name the folder for you. To change the name of the folder, tap into the *box* and use your keyboard to enter in a name.
5. Tap out of the folder to return to the home screen with your apps still shaking.
6. You can organize that folder the same way you can organize apps. You can move more apps into that folder by tapping, holding, and dragging apps into the folder. Perform a <u>home swipe</u> or tap <u>Done</u> at the upper right when you are finished.

How to Delete Apps (<u>Figure 15.3</u>)

1. With light pressure, tap and hold down on an app on your home screen until all of your apps start to shake, then release your finger.
2. To delete an app, tap the small (<u>x</u>) that appears at the upper left of an app.
3. Then tap <u>Delete</u>.
4. To delete a folder, just move all the apps inside a folder out of that folder.
5. Please note that some apps that came with the iPhone cannot be deleted.
6. Once an app is deleted, you can always get it back in the App Store, which is covered in the next chapter.
7. Perform a <u>home swipe</u> or tap <u>Done</u> at the upper right when you are finished.

Tap the small "x" to
delete the app

Drag and drop with
your finger one app to
inside another app to
create a folder on
your home screen.

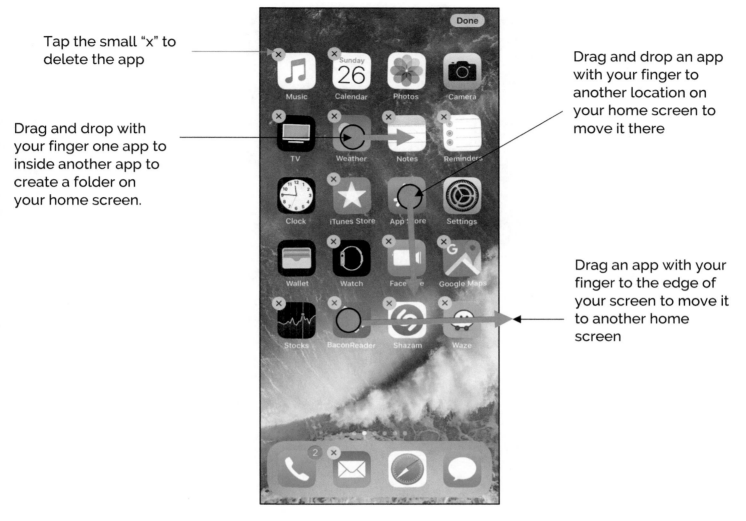

Drag and drop an app
with your finger to
another location on
your home screen to
move it there

Drag an app with your
finger to the edge of
your screen to move it
to another home
screen

Figure 15.3 – Managing & Organizing Apps

97

Chapter 16 – Apps

All the things that your iPhone can do are functions of apps. When you make a phone call, you are using the Phone app. When you send a text message, you are using the Messages app. When you browse the Internet, you are using the Safari app. Even when you are changing your settings, you are using the Settings app. The apps that come with the iPhone X are extraordinarily useful, but these are just a grain of sand in a desert compared to the vast amount of apps available. To get new apps, you will need to use an app called the App Store,

which is located somewhere on your home screen.

App Store

Let's open the App Store by tapping on it on our home screen. In order to use the App Store, you must have an Apple ID and you must be signed in to your Apple ID on your iPhone. If you still have not set up your Apple ID yet, I strongly urge you to do so. See Chapter 5 on setting up your Apple ID.

Browsing the App Store

Here in the App Store (Figure 16.1) you can browse through the massive library of apps available for download. Again, make sure you are signed in with your Apple ID, and make sure you remember your Apple ID password as you may need it to download apps.

Here at the main page of the App Store, called the Today page, you can see all the new and featured apps that Apple recommends. Let's take a look at the layout of this page, and the App Store itself, particularly the tabs at the bottom.

- **Today** – The Today tab shows you exactly what you see now, featured apps and content.
- **Games** – This tab allows you to browse gaming apps.
- **Apps** – This tab allows you to browse through all apps available.
- **Updates** – Shows you if any updates are available for apps.
- **Search** – Allows you to search for a specific app or function of an app.

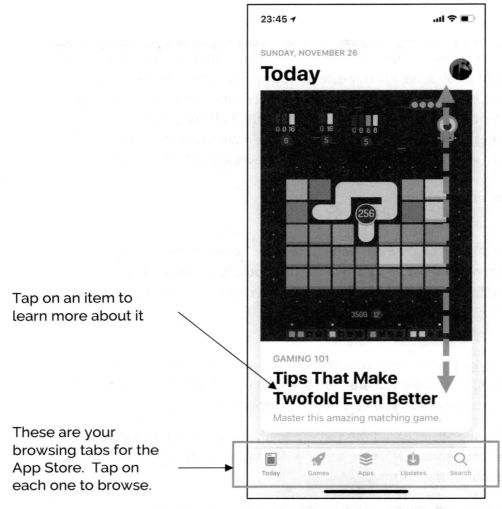

Tap on an item to learn more about it

These are your browsing tabs for the App Store. Tap on each one to browse.

Figure 16.1 – The App Store -> Today Tab

Browsing By Category

One of the best ways to find good apps is to browse by category. To do this, first tap the Apps tab at the bottom, then scroll down until you see Top Categories (Figure 16.2). Some categories will be shown and you can tap on one to explore it. For now, tap on See All to view all app categories.

You can tap on a category to see apps in that category, along with the top charts for that category.

Tap to view a list of all app categories

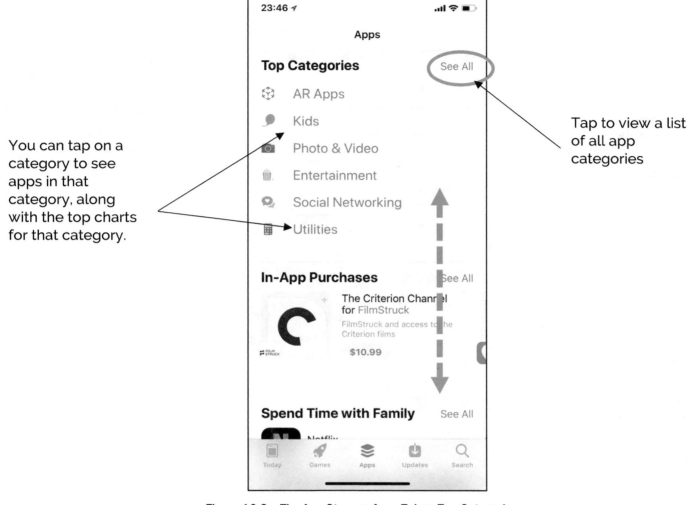

Figure 16.2 – The App Store -> Apps Tab -> Top Categories

When viewing all categories, you can scroll up and down to see the full list. Simply tap on a category to see apps within it. Within each category page, you will see Apps that are recommended by curators and the top charts for that category. The top charts include the most downloaded and used apps in that category and they are sorted by free apps and paid apps. Paid apps are apps you have to pay for in order to download. As always, you can use the back arrows at the upper left to go back.

Viewing an App

(See Figure 16.3) When you see an app that interests you, you can tap on the app itself to view more information. On this screen you will see the name of the app, the price to download it if there is one, and you will see the app's user ratings, rankings, age recommendation, and more. By scrolling down and scrolling left and right you can see screenshots of the app, read a description of the app, and read user ratings and reviews.

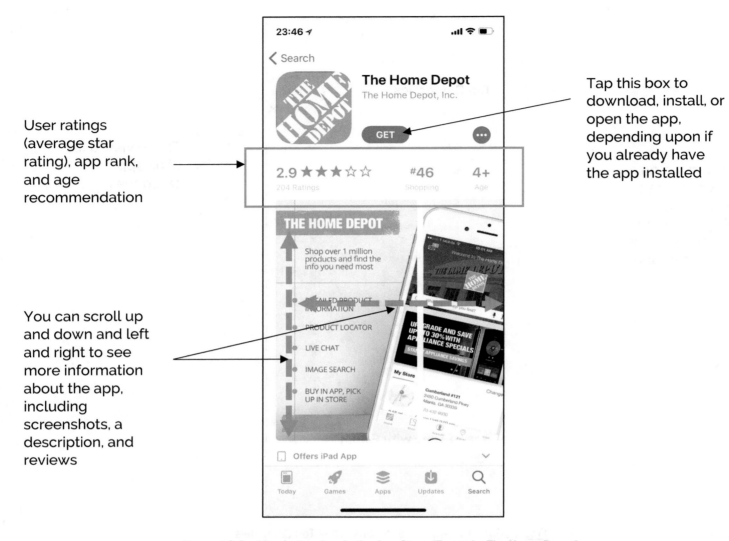

User ratings (average star rating), app rank, and age recommendation

Tap this box to download, install, or open the app, depending upon if you already have the app installed

You can scroll up and down and left and right to see more information about the app, including screenshots, a description, and reviews

Figure 16.3 – Viewing an App in the App Store (Example: The Home Depot)

Top Charts

(See Figure 16.4) Using Top Charts is one of my favorite ways to find popular apps. We have already demonstrated how to view Top Charts for categories, but you can also see Top Charts for all apps regardless of category. To access these Top Charts, first make sure you are on the Apps tab. Then, scroll down until you come upon the headlines **Top Paid** and **Top Free**. These two sections show you the most downloaded and used paid and free apps. You can tap on See All to see the full list, and you can tap on any app to learn more information.

Tap on an app to see more information

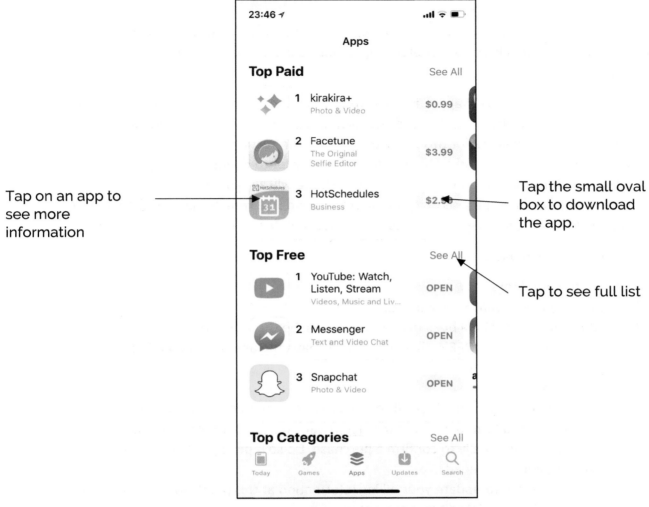

Tap the small oval box to download the app.

Tap to see full list

Figure 16.4 – The App Store -> Apps Tab -> Top Charts

Searching for Apps

If you know the specific app you want to download, you can search for it using the <u>Search</u> tab at the bottom. Tapping on this tab will allow you to search for any app by tapping into the <u>search bar</u> at the top of the screen. Simply type in the name of the app, or a subject such as "photo" and tap <u>search</u> on your keyboard. Usually, when you type the search term into the search field the App Store will make suggestions as to what you are looking for. You can tap on that suggestion to quickly enter it in and search. Now you will see a list returning your search results, and you can view these apps by tapping on them.

Updates

The <u>Updates</u> tab shows you what updates, if any, are available for apps you already own. You can tap <u>Update</u> next to the app to install the update. Alternatively, you can tap <u>Update All</u> at the upper right to install all available updates.

Downloading Apps

Now that we have seen how we can browse apps, let us see how we actually download one to our iPhone. The process is actually pretty simple, but before we begin let me give a note on your Apple ID.

As emphasized earlier in this chapter, in order to use the App Store you must be signed in with your Apple ID. You are also going to need your Apple ID password to download apps. On occasion, your iPhone will ask for this password just for verification purposes. Lastly, you will need credit card information linked to your Apple ID in order to download certain apps. Adding your credit card information is fairly simple. Your iPhone will ask you when it needs this information, and will bring up a screen with directions. Adding your credit card data does not mean you will be charged. Apple just needs it IN CASE you decide to purchase something.

Here is how you download an app:

1. Open the App Store.
2. Find the app you want to download.
3. Next to the app will be a box. Inside this box will be the words GET or a price such as $0.99. If the box says OPEN, then you already have the app downloaded. (See Figure 16.4)
4. Tap this box.
5. You may be asked for your Apple ID password at this point or Face ID. Enter it in.
6. You may be asked to double press the sleep/wake button in order to proceed with your download. This is usually to confirm a purchase. Do so if prompted and you want to go ahead with your download.
7. You may be asked to update your billing information at this point, such as credit card and billing address. Enter it in if prompted.
8. Your download will now commence. When it is finished, your newly downloaded app will appear on your home screen somewhere.

Using Downloaded Apps

When you download an app, it will appear on your home screen. To open it, just tap it. Using an app from the App Store is the same as any using any app that comes with the iPhone X. You can usually scroll through screens on the app, tap icons, and more. Each app is unique to itself, so you will have to play around with the app to get the hang of it.

There are certain aspects of apps that are important to consider. The first is in-app purchases. Many apps allow you to buy features or items from directly inside the app. These purchases could be access to content or even physical goods. To make these purchases you will most likely be using your Apple ID. Another important aspect is advertisements. Some apps have advertisements running through them. For most apps these are non-conspicuous and do not interfere with using the app, but for others the advertisements can be quite bothersome. If an app is flooded with advertisements and popups that make it difficult to even use the app, I would consider deleting the app and using something else. Apps like these are just trying to spam you with advertisements rather than provide a good service.

It is also worth noting that some apps offer a paid and free version. Usually the paid version comes with extra features or no advertisements.

Resources for Learning How to use Apps

You can check out the appendix of this text for a list of popular and useful apps. You might be amazed at what types of apps are out there and what they can do.

Some apps are a little tricky to learn, and there are a plethora of resources out there to help you. For learning how to use certain apps, we recommend checking out www.infinityguides.com. That website is dedicated to teaching people how to use many popular apps from a beginner's perspective.

Chapter 17 – Notifications

So far we have covered all the basics of using the iPhone X. You now know how to navigate the iPhone, perform a home swipe, make phone calls, exchange text messages, browse the Internet, use your email, personalize your device, download apps, and more. Now we are going to get into specific features that are absolutely essential to understand, and we start with Notifications.

Overview

Notifications are an integral part of your iPhone. Every time you receive a text, you will receive a notification. Every time you miss a phone call, you will receive a notification. In fact, every time certain information is delivered to you from an app, you will receive what is called a notification.

Notifications appear on your iPhone X in several forms. On your lock screen, you will see notifications lined up (Figure 17.1). While using your iPhone and it is unlocked, you will see notifications in real-time pop up at the top of your screen (Figure 17.2). When these notifications appear, you can tap on it to be brought directly to the notification. For instance, if you receive a text message while you are doing something on your iPhone, a box will appear at the top of your screen showing the text message and who it is from. This box will appear briefly. While the box is there, you can tap on it to be brought directly into the thread of the text message so you can respond. Alternatively, you can tap that box and drag your finger down to reply immediately.

Another type of notification is the badge icon, which is simply the red bubble you can see above an app on the home screen that has a number in it. This usually means you have a certain number of notifications available for that particular app. For instance, your Messages app will show a number in a red bubble above it on your home screen. This number indicates how many unread text messages you currently have.

Figure 17.1 – Notifications on the Lock Screen

Figure 17.2 – Notifications on Home Screens

Notifications can also appear in real-time as an alert. An alert is larger box that appears in the middle of your screen and they usually require you to dismiss the notification before it goes away. An example of this would be an emergency severe weather alert or something similar.

Notifications on the Lock Screen

When your iPhone X is locked and you look at notifications on your lock screen, you will notice that the notification presents the most basic information, such as that you received a text message from someone (Figure 17.1). If you have Face ID enabled and you look at your Notifications, they will change when Face ID recognizes you to show you more detail. Here is an example of this: if you receive a text message from John Doe and you glance at your lock screen you will see a notification that says: Messages, John Doe, Text Message. Now if you look directly at your iPhone's screen and you have Face ID enabled, after a moment Face ID will recognize your face and the notification will change to show you what the text message actually says. The same goes for other notifications as well.

The Notification Panel

At any time you can view all of your current notifications by accessing the Notification Panel. To do this, tap down at the very top of your screen at any time and swipe your finger down. This is your Notification Panel (Figure 17.3). Here you can see all the notifications that you have not acknowledged. To go to a particular notification, simply tap on it. To clear out a section of notifications, tap the small x next to the date, and then tap clear. Just perform a home swipe to return to your home screen.

To open the Notification Panel, tap down at the very top of your screen and swipe down

Tap and swipe to the right to quickly open the notification (or just tap on the notification)

Tap to clear the particular set of notifications

Tap and swipe to the left to view the complete notification or clear it. Swiping completely to the left will clear the notification.

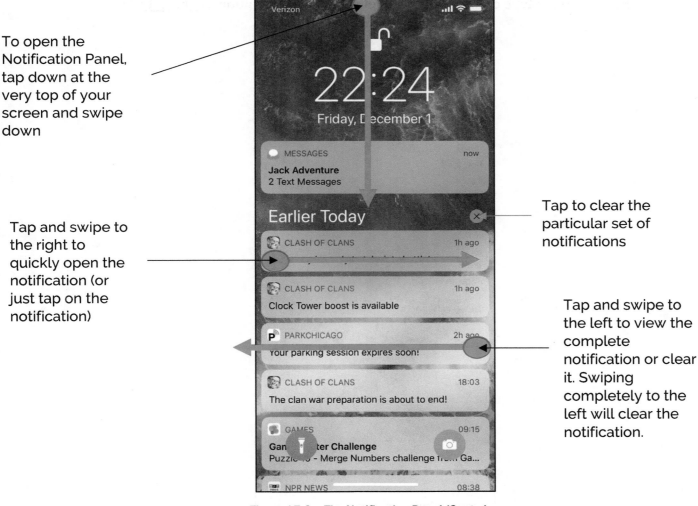

Figure 17.3 – The Notification Panel (Center)

Examples of Notifications

As stated earlier, a notification can be from any app. For instance, a breaking news story from Fox News can be a notification if you have the Fox News app. A new Facebook like on your post can be a notification if you have the Facebook app installed. Basically, any app can deliver notifications.

Managing Notifications

Sometimes, you may not want to get notifications from a certain app. These can be easily managed in the Settings app.

1. Open the <u>Settings</u> app.
2. Tap <u>Notifications</u>.
3. Tap the app you want to manage notifications for.
4. Now you can choose whether you want to allow notifications from this app. You can also choose the type of notification as well as some other options. (<u>Figure 17.4</u>)

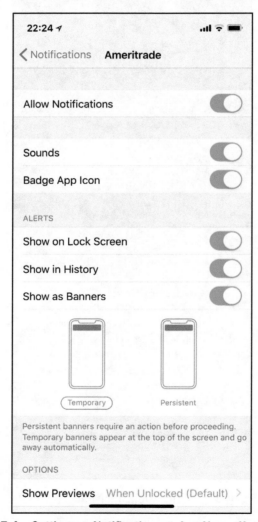

<u>Figure 17.4</u> – Settings -> Notifications -> App Name (Ameritrade)

Chapter 18 – The Control Center

The Control Center, sometimes referred to as the "control panel," is a nifty tool on the iPhone X that allows you to quickly change some settings. Let's explore it.

To Access the Control Center

1. On any screen on your iPhone, tap down at <u>the top right corner of your screen</u> and swipe straight down. This will bring up the Control Center (<u>Figure 18.1</u>).

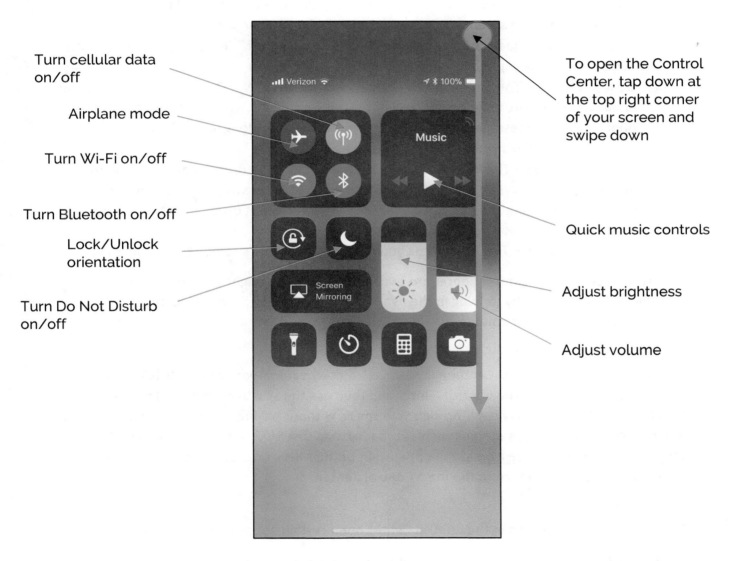

<u>Figure 18.1</u> – The Control Center

Control Center Functions

Inside your Control Center you will see some buttons. Tapping these buttons turns the selected setting on or off.

- **Airplane Mode** – Tapping on the airplane button turns airplane mode on or off. When a button is highlighted that means it is currently enabled. Airplane mode turns off all communication abilities on the iPhone, including cellular, Wi-Fi, and Bluetooth. Once you turn Airplane mode on, you can still enable Wi-Fi by using the Control Center.
- **Wi-Fi** – The radio waves is the Wi-Fi button. Tapping this will turn Wi-Fi on or off.
- **Cellular Data** – The antenna icon enables or disables cellular data.
- **Bluetooth** – The Bluetooth button turns Bluetooth on or off. To connect to a Bluetooth device, you need to go to Settings -> Bluetooth.
- **Do Not Disturb** – The moon button enables or disables Do Not Disturb mode. When you enable Do Not Disturb mode, you will not receive notifications or hear any sounds for phone calls and messages when your screen is off. These along with all other notifications besides your alarm clock will be silenced.
- **Lock Orientation** – The lock orientation button will lock the orientation of your iPhone in portrait mode. When this is enabled, turning your iPhone to landscape orientation will not re-orientate the screen. If your iPhone screen is not orientating to landscape mode this option may be on.
- **Screen Mirroring** – This button lets you mirror your iPhone's screen to an Apple TV.

On the right side of the Control Center are two vertical ovals for brightness and volume. You can adjust these settings by tapping in the oval and dragging up or down to increase or decrease the brightness or volume.

Music Controls

The Music box in the Control Center allows you to quickly control your music. While a track is playing, you can bring up the Control Center and then use the skip back, pause/play, and skip forward icons to quickly navigate through your music without having to open up the Music app. You can also bring up more options such as broadcasting to a Bluetooth connected speaker by tapping on the sound wave icon at the upper right of the Music box. Simply tap out of the enlarged Music box to return to the main Control Center.

Control Center Apps

The bottom row of buttons in the Control Center open certain apps or perform functions.

- **Flashlight** – Tapping on the flashlight icon turns on your iPhone's flashlight, which is just the flash on your camera. Tap the flashlight again to turn it off. This function is very useful.
- **Timer/Alarm Clock** – The clock icon opens the Clock app, and brings you directly to the Timer tab. In the Timer tab, you can set a timer that will notify you when the desired

time elapses with an alert and sound. You can also set your alarm clock from here by tapping on the <u>Alarm</u> tab at the bottom. To create a new alarm, tap the <u>plus icon</u> at the upper right. Alternatively, you can enable a previously created alarm by tapping the <u>small oval</u> next to a time. You can now set the time for your alarm and specify which type of alert you want. Tap <u>Save</u> at the upper right to finish setting the alarm.

- **Calculator** – The third icon at the bottom is the Calculator app. Tap this to open up the calculator, which you can use.
- **Camera** – The last icon at the bottom opens the Camera app.

3D Touch Functions in the Control Center

You can use 3D Touch on many of the Control Center buttons. As a reminder, 3D Touch is pressing down on the screen with a little additional force and holding it for a moment. Two 3D Touch examples are worth noting in the Control Center:

- **Night Shift** – To enable Night Shift, press on the <u>vertical brightness adjustor</u> with a little additional force and hold until the adjustor comes to full screen. Underneath the adjustor will be the <u>Night Shift</u> button, which you can tap on to enable. When you turn Night Shift on, your iPhone will change its color display. The purpose of Night Shift is for when you are using your iPhone late at night and do not want your iPhone's screen straining your eyes and potentially keeping you awake. Night Shift will automatically turn off the next morning.
- **Flashlight** – Performing 3D Touch on the <u>flashlight button</u> allows you to adjust the power of the flashlight. Simply tap or drag your finger on the scale to adjust the power.

Exiting the Control Center

To exit the Control Center either perform a <u>home swipe</u> or tap down on the Control Center and swipe it back up.

Adding Buttons to the Control Center

You can add or remove which buttons appear in the bottom row of the Control Center, and even create several rows of buttons. Here's how:

1. Open the <u>Settings</u> app
2. Tap <u>Control Center</u>
3. Tap <u>Customize Controls</u>
4. On the following screen, you can remove buttons by tapping the <u>red minus symbol</u> next to a function and add buttons by tapping the <u>green plus symbol</u>

Chapter 19 – Siri

Siri, as you may have heard, is Apple's intelligent voice assistant. You can talk to Siri, and Siri will listen to what you say and try to do whatever you ask. The possibilities of what you can do with Siri are endless.

Turning Siri On

By default, Siri is enabled on your iPhone. To be sure, let's make sure Siri is enabled. To do this, follow these steps (Figure 19.1):

1. Open Settings.
2. Find and tap Siri & Search.
3. Make sure Press Side Button for Siri is enabled at the top. (The slider tab next to it should be green. If not, tap it.)

While we are in the Siri settings screen, let's take a quick look at some of the settings you can adjust.

- **Allow Siri When Locked** – Allows you to access Siri when your iPhone is locked. I usually leave this enabled so I can access Siri quickly without unlocking my iPhone.
- **Listen for "Hey Siri"** – With this enabled, you can access Siri by simply saying "Hey Siri" at any time. If your vehicle has Apple CarPlay, you will need this enabled.
- **Other Settings** – The settings at the bottom allow you to change some other settings including which language you want Siri to use and which voice you want Siri to have.

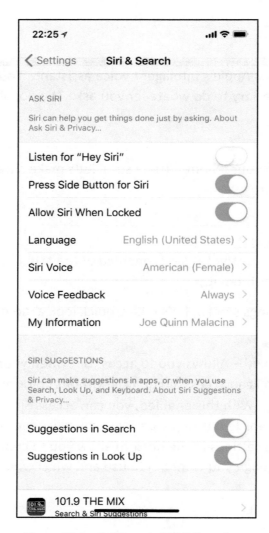

Figure 19.1 – Settings -> Siri & Search

Accessing Siri

Now that Siri is enabled on your iPhone, let's go back to the home screen. From the home screen accessing Siri is very easy. The easiest way to access Siri is to press and hold the sleep/wake button until you see a sound wave appear at the bottom of your screen (Figure 19.2).

You can make Siri start listening again by pressing on the sound wave (or this icon shown), or pressing and holding the sleep/wake button again

Figure 19.2 – Using Siri

Using Siri

Now let's try some examples to see exactly how Siri works. First, let me note that when you access Siri by holding the sleep/wake (power) button, you have two different approaches. One is, when the sound wave appears, you can let go of the power button and start speaking. Two is, when the sound wave appears, you can continue holding down on the power button, start speaking, and release the power button when you are done speaking. Either way works. For simplicity, I am going to use the former method in our examples.

So let's press and hold the power button until the sound wave appears then release the power button. Now we can say, "What is the weather?" Siri will listen, then respond with some information. She can either speak the information back to you or bring up a screen with the information you requested. To start a new Siri session or attempt to speak again, tap on the graphic that appears at the bottom of your screen or just hold the power button again.

Let's try another example, press and hold the <u>power button</u> and release, and then say "Show me my Photos." Siri will listen and then automatically open the Photos app for you.

There are countless things you can say to Siri that she will perform. Try out anything you'd like. A few more examples are listed below, and you can see a much larger list of examples in the appendix of this text.

- Make Phone Calls – "Call *Contact Name*"
- Lookup Contact Information – "Lookup *Contact Name*"
- Lookup Sports Scores – "Who won the White Sox game today?"
- Set an Alarm Clock – "Set an alarm for 6:30 AM"
- Do Math – "What is two plus two?"
- Visit Websites – "Bring me to Infinity Guides dot com"

Chapter 20 – Native Apps

Let's explore some native apps that come preloaded on the iPhone. Many of these are very useful and you may find yourself using them often.

Music

The Music app is where you can play and listen to your music. There are three ways to listen to music. The first is to load songs that you already own onto your iPhone. To do this, you will need to use iTunes, which is a program for computers. To see how to do this, go to www.applevideoguides.com and click on iPod and iTunes Guide at the top.

The second way to listen to music is to use the Apple Music service. This service lets you listen to nearly any song you want at any time, right from your iPhone and any computer or device. The Apple Music service is huge, and is a subscription service from Apple. To learn how to use Apple Music, go to www.applevideoguides.com/music.html.

The third way is to purchase and download music from the iTunes Store app on your iPhone or through iTunes on a computer. Once you purchase and download some music, it will automatically appear in the Music app.

Once you have music on your iPhone using any of the two above methods, using the Music app is pretty simple. You can browse through the app using the tabs at the bottom, and simply tap on a song to play it. You can also quickly skip through tracks in the Control Center, which is covered in Chapter 18.

Maps

The Maps app is a great app for turn-by-turn navigation. In other words, it is a GPS right on your iPhone (Figure 20.1). Inside the Maps app you can tap into the search bar and type in the address of where you want to go. You can also search for nearby places. While typing an address or place, Maps will present suggestions as to what you are looking for. If yours comes up, you can tap on it. Tap on Directions to get driving directions to the location. Your iPhone will give you turn-by-turn directions as you are driving. Furthermore, the Maps app works great with vehicles that have Apple CarPlay.

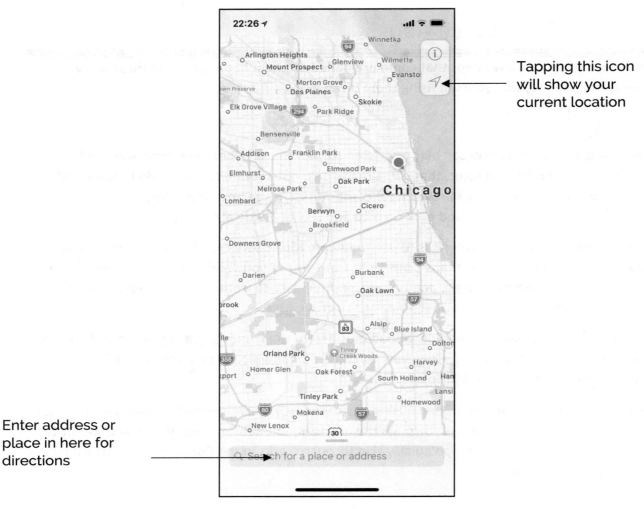

Figure 20.1 – Maps App

Note: There are other apps that offer GPS and turn-by-turn directions. Two popular ones are Google Maps and Waze, available in the App Store for free.

Weather

The Weather app is a very simple app that shows you the current weather and forecast. By default, the Weather app will show you your current location's weather, as long as you have Location Services enabled in Settings.

To add an area to the Weather app, tap the 3 lines at the bottom right. Now tap on the plus sign and then you can enter a postal code to quickly add that area to your Weather app. You can browse through your different saved locations by swiping left or right. You can also scroll up and down to see additional weather information. (Figure 20.2)

Tap here to add
another location to
check weather

Figure 20.2 – Weather App

FaceTime
(See Figure 20.3)

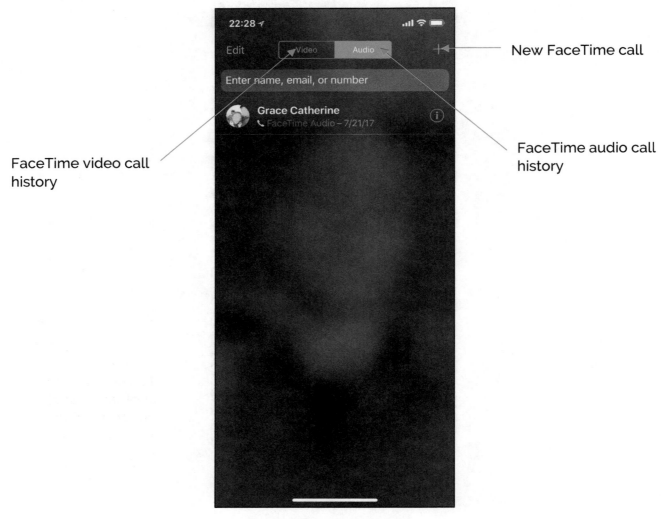

FaceTime video call history

New FaceTime call

FaceTime audio call history

Figure 20.3 – FaceTime App

FaceTime, which I have mentioned briefly earlier in this text, is an app that allows you to partake in video calls with other Apple users. This app is very similar to Skype, which is also available for free in the App Store. When you open FaceTime, you can place a video call by tapping on the plus sign at the upper right. Now you can look through your contacts to find someone to video chat with. When you have found that person, tap on their *name*. Now FaceTime will determine if they are able to make a FaceTime call. You will know if you are able to FaceTime that person because some icons may appear next to their name. If the video recorder icon appears, that means that you can make a FaceTime video call with them. If the phone icon appears next to their name that means you can make a FaceTime audio call with them, which has no video. If both icons are greyed out, and you are not allowed to tap on them, that means you cannot make a FaceTime call of any type with them. In other words, they most likely do not have FaceTime on their device. (Figure 20.4)

FaceTime calls have pretty much the same controls as regular phone calls. You can change which camera you are using (selfie or regular) by tapping on the camera icon during the call.

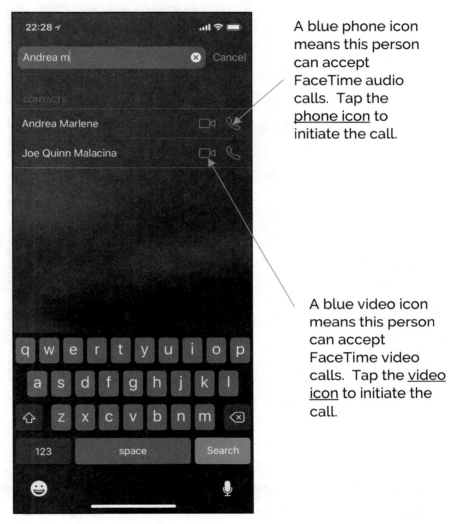

A blue phone icon means this person can accept FaceTime audio calls. Tap the phone icon to initiate the call.

A blue video icon means this person can accept FaceTime video calls. Tap the video icon to initiate the call.

Figure 20.4 – FaceTime Availability

Clock

We have already covered some of the Clock app in Chapter 18. There are a few things we missed however, so we will cover them now. In this app you can see the time around the world, as well as set an alarm, and start a timer. You can also use the stopwatch function by tapping on the Stopwatch tab at the bottom. Lastly, the Bedtime tab allows you to set up a custom bedtime and wake-up alarm, depending upon how you sleep. (Figure 20.5)

Tap to edit previous alarms

Create a new alarm

Tap to enable a previous alarm

Clock tabs for different functions (alarm clock, bedtime function, stopwatch, timer, and world clock)

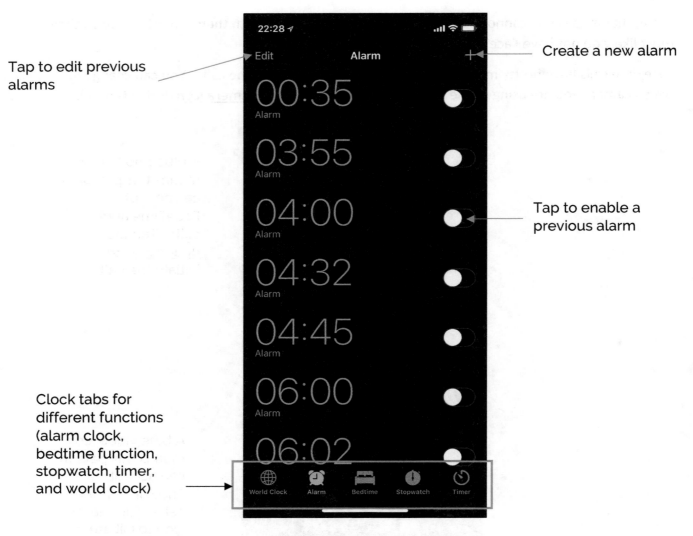

Figure 20.5 – Clock App (Alarm Tab)

Notes

The Notes app is an incredibly useful app for creating notes. It is very easy to use; just open the Notes app, tap the pencil and square icon at the lower right to start a new note, and begin typing away. Notes save automatically, and you can access all of your saved notes in the main notes screen (use the back arrows at the upper left).   You can delete notes by tapping on Edit at the upper right.

Furthermore, while creating a note you can sketch and use additional tools by using the icons at the bottom of your screen. You can also password protect certain notes by opening the note, tapping the rectangle with an arrow inside it icon at the upper right, and then tapping lock note. Notes save automatically. (Figure 20.6)

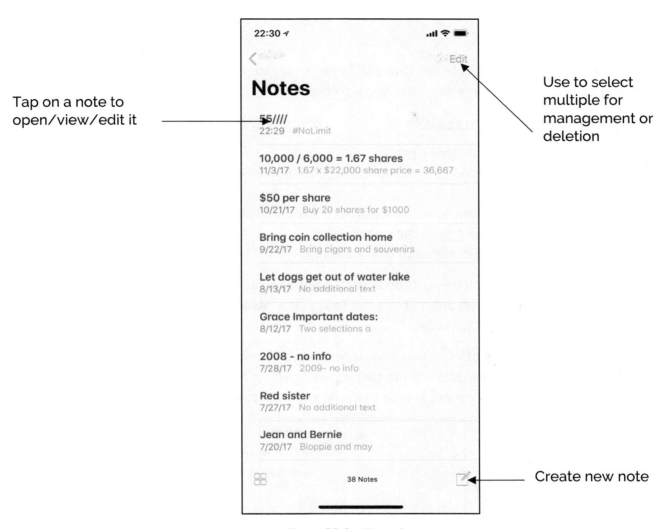

Tap on a note to open/view/edit it

Use to select multiple for management or deletion

Create new note

Figure 20.6 – Notes App

iBooks

iBooks is the app that lets you read books on your iPhone. The iPhone screen is a little small, so this app may be better used on an iPad. All electronic books that you download and store in iCloud or place in your device through iTunes will be available in iBooks.

You can also browse through books and download them using the tabs at the bottom of iBooks, including the Search and Top Charts tab. You will need to be signed in with your Apple ID to purchase and download books. Once you purchase a book using your Apple ID, it will automatically be sent to all of your Apple devices that are signed in with your Apple ID. So if you buy a book on your iPhone, it will also appear on your iPad for easy reading. You may even be reading this book on your iPad!

Calendar

The Calendar app is a pretty simple app that allows you to create events and reminders. To view all the events you have on a day, simply tap on that day.

To Create a New Event (See Figure 20.7)

1. Tap the plus icon at the upper right.
2. Enter in all the corresponding information, if required. Include the date and time (if applicable).
3. Choose whether you want to set an alarm to remind you about this event.
4. When you are done, tap Add at the upper right.
5. Your event has now been added to your calendar.

You can tell if you have scheduled events on a day if a dot appears below the date. You can tap on Today at the lower right to be brought to your Today view, which shows you what events you have coming up, starting with today.

You can also see a quick view of what you have coming up on your calendar from the home screen. Simply go to your home screen, and swipe your finger to the right until you get to the Today screen. If you have any events in your calendar coming up, they will appear here.

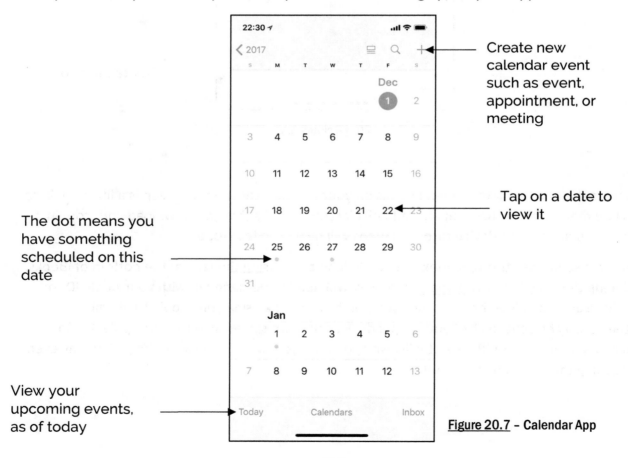

Create new calendar event such as event, appointment, or meeting

Tap on a date to view it

The dot means you have something scheduled on this date

View your upcoming events, as of today

Figure 20.7 – Calendar App

Chapter 21 – Tips & Tricks

Congratulations! You have made it through the majority of this book, and you should now have a solid understanding of how to use your iPhone X. There are no more "basics" left to teach you, so I will leave you with a few tips and tricks that you may find helpful when using your iPhone.

Backing up your Device

This is a MAJOR tip. You should always have a backup of your iPhone's data available in case anything happens. If you destroy or lose your iPhone, a backup will allow you to get all of your data back. There are two ways to back up your device.

1. Backup to iCloud (Recommended)

Backing up to iCloud is definitely the way to go. This is done automatically and continuously so you never have to worry about anything. You will need an Apple ID to do so, and you must be signed in to iCloud with that Apple ID (covered in Chapter 5). To turn iCloud backup on, follow these steps:

1. Open Settings.
2. Tap the big box at the top with your name.
3. Tap iCloud.
4. Tap iCloud Backup.
5. Now make sure iCloud Backup is on (enabled green). If it was off previously, tap on Back Up Now to begin the backup.

When your iPhone is backing up to iCloud, you do not need to worry about setting up backup again, as your iPhone will backup periodically, usually when you are sleeping and your iPhone is charging. Please note your iPhone must be connected to Wi-Fi in order to initiate its automatic backup. Now should anything ever happen to your iPhone, you can always get your data back. iCloud saves your data securely in the cloud.

2. Backup to iTunes

The second way to backup your device is to use iTunes. Doing this saves a backup on your computer that can quickly be restored. To learn how to do this, visit www.applevideoguides.com and click on iPod and iTunes Guide at the top.

Automatically Update Apps

Turning on automatic updates for apps is a great time-saving feature of the iPhone. When this feature is on, you will not need to ever manually update apps through the App Store. Here are the steps:

1. Open Settings
2. Tap iTunes & App Store

3. Tap the button icon next to <u>Updates</u> under Automatic Downloads to enable it

Taking a Screenshot

A screenshot is a picture of exactly what your screen looks like, and can be very useful. For instance, say you received a text with instructions on how to bake a cake, and you want to share those instructions with a friend via text message. One way to accomplish this is take a screenshot of the original text and send the picture of that screenshot to your friend.

At any time you can take a screenshot of what you are viewing on your iPhone X by pressing the <u>power button</u> and <u>volume up button</u> at the same time and then release. You will know a screenshot was taken when you see your screen flash, and the screenshot appears in the lower left corner of your screen. You can tap on the picture to view it, and several sharing and editing features will become available to you. Screenshots will save as pictures in your Photos app.

Background Apps & the App Switcher

(See <u>Figure 21.1</u>) Whenever you open an app, it remains open in the background even after you leave the app. It can be beneficial to completely close an app running in the background if it is not working properly. You can also use this method to quickly go to an app you were recently using.

To close down a certain background app, first tap and hold on any app in the App Switcher until the minus symbols appear at the top left corner of each app screen.

To access the App Switcher, perform a home swipe BUT do not lift your finger off the screen. Instead, leave your finger on your screen at the end of the home swipe until app screens appear at the center of your screen. Then you can release your finger.

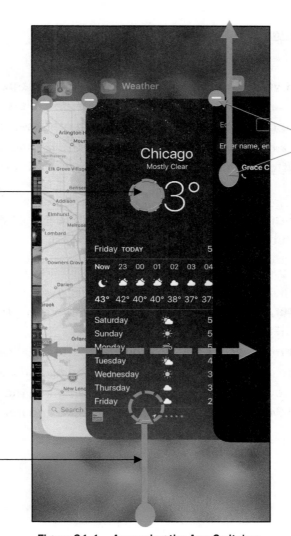

After you have performed the step to bring up the minus symbols, you can close an app down by tapping on the minus symbol or by tapping on an app screen and swiping it up and off the screen. You can even swipe off every background app running for a "clean slate".

Figure 21.1 – Accessing the App Switcher

How to Close Background Apps OR Go to a Recent App

1. Perform a home swipe but DO NOT lift your finger off the screen at the end. Instead, leave your finger on your screen near the middle until several app screens appear of apps you have recently used. You can now release your finger. Your home screen should break away and you should see numerous screens next to each other. (If it did not work, try again. This gesture can require some practice.)
2. If you want to quickly go to another app you used recently, swipe left or right and tap on the screen of the app you want to go to.
3. To close down an app, tap down and hold on the screen of any app until red minus symbols appear at the upper left corner of each background app. Now to close an app down, tap on any background app and swipe it up and off the screen. This closes down an app completely.
4. Alternatively, you can tap on the red minus symbol to close down a background app.

5. Perform a <u>home swipe</u> to return when done.

Restart your iPhone

It is best practice to restart your iPhone every now and then, the same way you would restart a computer to keep it fresh. I recommend restarting your iPhone if it is running slow or not working properly. Doing so can fix the problem. As a good practice, I recommend restarting your iPhone at least twice a month. To do so, simply turn the iPhone off by holding the <u>power button</u> and <u>any volume button</u>. Wait 60 seconds after the phone is off, then turn it back on.

3D Touch on Everything

In Chapter 15, we covered the 3D touch feature of the iPhone. Now I will show you a couple of ways you can use 3D Touch that may not have occurred to you.

1. In the <u>Messages app</u>, you can tap and hold with some pressure to bring up a quick preview of a <u>thread</u>, and quick reply options. (<u>Figure 21.2</u>)
2. In the Control Center, you can tap and hold with some pressure on the <u>Timer icon</u> to quickly set a timer. (<u>Figure 21.3</u>)
3. When typing a text message, or any text for that matter, you can quickly move your text marker to anywhere in the text message box by tapping and holding with some pressure on the <u>keyboard</u>. Now, while holding down, you can move the marker wherever you wish. This is useful when you have written a long text, and need to edit very specific portions of it.

Figure 21.2 – Messages App: 3D Touch on a Thread

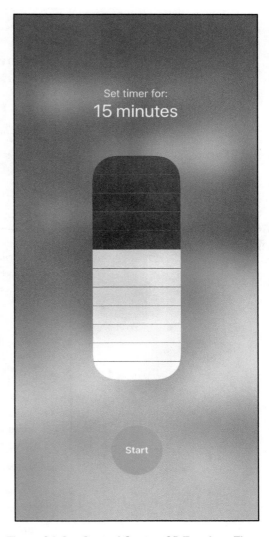

Figure 21.3 – Control Center: 3D Touch on Timer Icon

Copy and Paste

We have already covered how to copy a picture. As a reminder, you can do this by using 3D Touch on a photo or photo thumbnail and swiping up, and then tapping Copy. To paste it in text somewhere, double tap with your finger in the text area, and then tap on Paste.

To copy and paste text, find the text you want to copy and tap and hold on it. Now move the small marker dots left and right by tapping and dragging them to select the exact portion of text you want to copy. Now tap the Copy box. You can paste the text by double tapping in a text box and then tapping Paste. (Figure 21.4)

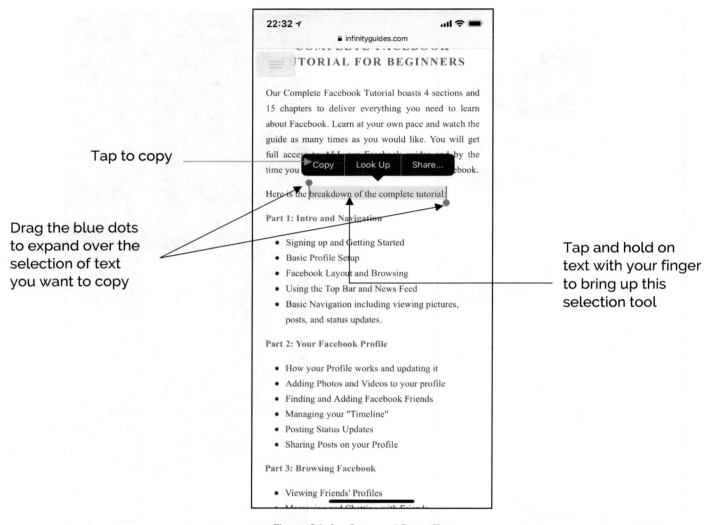

Figure 21.4 – Copy and Paste Text

Low Power Mode

There are some things you can do to improve the battery life of your iPhone. The biggest way to preserve your battery life is by switching on Low Power Mode, which reduces power consumption temporarily. Low Power Mode is an excellent feature to use when your battery is running low and you cannot afford for your battery to die anytime soon. To enable it, follow these steps:

1. Open Settings
2. Tap Battery
3. Tap the enable box next to Low Power Mode to enable it
4. To turn it back off tap it again

Personal Hotspot

Your iPhone X has the ability to become a personal hotspot, which in layman's terms is a device which can broadcast its internet connection to other devices. Using the personal hotspot feature is limited by your cellular provider, and some providers charge to use this service. So before you enable your personal hotspot, check with your cellular provider.

To enable the personal hotspot:

1. Open Settings.
2. Tap Personal Hotspot.
3. Enable Personal Hotspot by tapping on the tabular icon next to Personal Hotspot. When the tab is green personal hotspot is enabled.
4. Set a Wi-Fi password for devices to access your personal hotspot by tapping in the Wi-Fi Password box and creating a password.

A personal hotspot is a great way to create a Wi-Fi network for your other devices when Wi-Fi is not available. Be sure to turn Personal Hotspot off when you are done, as it uses significant battery power and data.

Siri Easter Eggs

As mentioned in the Siri chapter, you can say anything you want to Siri. Anything. You may even discover some Easter eggs while speaking to her. Try these: (Press and hold the power button to begin).

- How much wood could a woodchuck chuck is a woodchuck could chuck wood?
- What do you look like?
- I love you.
- I love my iPhone.
- I am tired.
- How are you doing today Siri?

Try some more yourself.

Chapter 22 – More Resources

This guide has covered all the beginner aspects of the iPhone. We have also covered many intermediate and advanced aspects, but there is still plenty more you can learn. Most notably, there is a lot to learn about specific apps. You can also learn more about Apple Music, iTunes, and social media on your iPhone.

- **Apple Music Learning** – www.applevideoguides.com/music.html
- **iTunes Software Learning** – www.applevideoguides.com/ipod-itunes.html
- **iPhone Apps Learning** – www.infinityguides.com
- **Facebook Learning** – www.infinityguides.com/facebook.html
- **Twitter Learning** – www.infinityguides.com/twitter.html
- **More Manuals for Beginners** – www.beginnermanuals.com

CONCLUSION

Thank you for taking the time to read this book. It is my hope that you feel much better about using your iPhone X. I am confident that if you took the time to read this entire book, then you will have no problem using every aspect of your iPhone with ease. Continue to use this book as a reference when you need it. The table of contents can quickly lead you to your answer, and the appendixes can be especially helpful as well and I hope you use them.

I welcome your thoughts and feedback on this text, please come visit my Facebook page online at www.facebook.com/joemal. You can also tweet me @JoeMalacina on Twitter. I am often online answering questions from people who have read this text, and helping people with complex iPhone issues. As a last piece of advice, please remember your Apple ID, Apple ID password, and lock screen passcode. Forgetting even one of these, especially your lock screen passcode, can be a real headache.

Enjoy using your iPhone X.

APPENDIX A – Additional Recommended Apps

- 560 The Answer – Talk Radio
- AirBnB – Travel
- Your Favorite Airline's App - Travel
- Amazon – Shopping
- Bible – Reading
- Brave – Web Browser
- Crossword – Game
- Daily Celebrity Crossword - Game
- DoodleJump - Game
- Fox Sports – Sports
- Expedia – Travel
- Facebook – Social Media
- Facebook Messenger – Social Media/Chat
- Fandango – Movies
- Flixster – Movies
- Fox News – News
- Google Maps – For Directions & GPS
- Groupon – Shopping
- Grubhub – Food & Dining
- HD Wallpapers – Wallpapers
- Heads Up! - Game
- The Home Depot – Shopping & Home Improvement
- iHeartRadio – Radio
- Instagram – Social Media
- Lyft – Rideshare Travel
- McDonalds App - Dining
- Microsoft Office – Productivity
- MLB App - Sports
- MyFitnessPal – Diet and Fitness
- Netflix – Entertainment
- NFL App - Sports
- OpenTable – Dining

- Pages – Word Processor
- Periscope – Social Media
- Radio.com – Radio
- Shazam – Music ID
- Shopular – Shopping
- SitorSquat – Miscellaneous
- Sky Guide – Miscellaneous
- Skype – For Video Chat
- Snapchat – Social Media
- Speedtest – Utilities
- SpotHero – Parking & Travel
- Target – Shopping
- The Weather Channel – Weather
- Twitter – Social Media
- Uber – Rideshare Travel
- UPS – Postal Tracking & Shipping
- VLC – Video Player
- Wall Street Journal – News
- Waze – For Directions & GPS
- WebMD – Health
- WhatsApp – Texting App
- Words with Friends - Game
- Yelp – Reviews
- Your Bank's App

APPENDIX B – Siri Examples

These are all examples of things you can say to Siri.

- What time is it?
- Show me restaurants around here
- Call *Contact Name*
- Send a text to *contact name*
- Open *app name*
- Check my email
- Check my messages
- Do I have any appointments today?
- Check my battery life
- Send a text to *contact name*
- Tell *contact name* I am on my way in a text
- What movies are playing?
- Play *song name*
- Play *playlist name*
- Play *artist name*
- Remind me to wash the dishes tonight
- Email *contact name* about the trip
- Wake me up at 7 AM tomorrow.
- Note that I need to get my dog a new bone
- Turn on Airplane mode
- Decrease/Increase my brightness
- What is Apple's stock price?
- What is the address of *contact name*?
- Change my wallpaper
- Play a random song
- Where can I get a burger around here?

APPENDIX C – List of Common Functions

- **Add New Email to iPhone:** Settings -> Accounts & Passwords -> Add Account
- **Change Language:** Settings -> General -> Language & Region
- **Change Lock Screen Password:** Settings -> Face ID & Passcode
- **Change Ringtones:** Settings -> Sounds & Haptics
- **Change Wallpaper:** Settings -> Wallpapers
- **Check for iOS Updates:** Settings -> General -> Software Update
- **Completely Restore iPhone to Factory Default (WARNING: this will delete all of your iPhone data and bring it back to right out of the box status. DO NOT DO THIS unless you know what you are doing.):** Settings -> General -> Reset -> Erase all Content and Settings - > enter passwords
- **Connect to a Bluetooth Device:** Settings -> Bluetooth -> My Devices
- **Connect to a Wi-Fi Network:** Settings -> Wi-Fi -> Tap on Network -> Enter password -> Tap Join
- **Create New Text:** Messages -> Square and Pencil Icon
- **Join Apple Music:** Settings -> Music -> Join Apple Music
- **Set Default Search Engine:** Settings -> Safari -> Search Engine
- **Clear Your Internet History:** Settings -> Safari -> Clear History and Website Data
- **Shut Down iPhone X (Turn Off):** Settings -> General -> Shut Down
- **Sign in to your Apple ID:** Settings -> iTunes & App Store -> Apple ID
- **Sign Into Facebook (requires app first):** Settings -> Facebook -> Sign In
- **Switch to Military Time:** Settings -> General -> Date & Time -> 24 Hour Time
- **Turn Location Services On/Off:** Settings -> Privacy -> Location Services
- **Turn on Do Not Disturb:** Control Panel -> Moon Icon
- **Turn on iCloud Photo Library:** Settings -> Photos & Camera -> iCloud Photo Library
- **Turn Siri On/Off:** Settings -> Siri
- **Create a new E-mail:** Mail -> Square and Pencil Icon
- **Delete a Recent Call History:** Phone -> Recents Tab -> Swipe left over a recent call
- **Add Emoji's to Keyboard:** Settings -> General -> Keyboard -> Keyboards -> Add New Keyboard... -> Emoji